Lecture Notes in Computer Science 7356

Commenced Publication in 1973
Founding and Former Series Editors:
Gerhard Goos, Juris Hartmanis, and Jan van Leeuwen

Michail Salampasis   Birger Larsen (Eds.)

Multidisciplinary Information Retrieval

5th Information Retrieval Facility Conference
IRFC 2012
Vienna, Austria, July 2-3, 2012
Proceedings

 Springer

Volume Editors

Michail Salampasis
Vienna University of Technology
Institute of Software Technology
and Interactive Systems
Favoritenstr. 9-11/188
1040 Vienna, Austria
E-mail: salampasis@ifs.tuwien.ac.at

Birger Larsen
Royal School of Library
and Information Science
2300 Copenhagen, Denmark
E-mail: blar@iva.dk

The conference is co-organised by
the Multilingual and Multifaceted Interactive Information Access
COST Action and the Vienna University of Technology,
Information Management and Preservation Group

ISSN 0302-9743   e-ISSN 1611-3349
ISBN 978-3-642-31273-1   e-ISBN 978-3-642-31274-8
DOI 10.1007/978-3-642-31274-8

Springer Heidelberg Dordrecht London New York

Library of Congress Control Number: 2012939842

CR Subject Classification (1998): H.3.3, H.3, H.2.8, I.2.7, H.4, H.5

LNCS Sublibrary: SL 3 – Information Systems and Application,
incl. Internet/Web and HCI

Typesetting: Camera-ready by author, data conversion by Scientific Publishing Services, Chennai, India

Printed on acid-free paper

Springer is part of Springer Science+Business Media (www.springer.com)

Preface

These proceedings contain the refereed papers presented at the 5th Information Retrieval Facility Conference (IRFC) for Science and Industry, which was held in Vienna during July 2–3, 2012. The conference was organized by the IC1002 COST Action Multilingual and Multifaceted Interactive Information Access (MUMIA) and the Vienna University of Technology.

COST is an intergovernmental framework for European Cooperation in Science and Technology, allowing the coordination of nationally funded research on a European level. The IC1002 COST Action on Multilingual and Multifaceted Interactive Information Access (MUMIA) coordinates collaboration between researchers and professionals working on three key disciplines for the future of search: machine translation (MT), information retrieval (IR) and multifaceted interactive information access (MIIA). To form a common basis for collaboration the domain of patent retrieval has been selected as a use case, as it provides highly sophisticated and information intensive search tasks. The MUMIA action aims to empower the synergies from the disparate research fields of MT/IR/MIIA within the specific context of patent search and other next-generation Web applications. More than 60 experienced scientists and young researchers from 26 countries participate in MUMIA's research network.

The 5th IRFC aimed to tackle three complementary research areas: information retrieval, machine translation for search solutions and interactive information access.

The IRF Conference provides a multi-disciplinary, scientific forum that aims to bring young researchers into contact with industry at an early stage. Another emphasis is put on the applicability of IR solutions to real industry cases and the respective challenges. A panel of experts from both academia and industry reviewed the relevance of the submitted papers. IRFC 2012 received 17 high-quality submissions, of which 11 were accepted for presentation at the conference and appear here. Each submission was reviewed by at least three Program Committee members. Five of these are IR papers in the patent search domain, two deal with the machine translation and natural language processing for the IR area, while the remaining four deal with Web IR and associated tasks.

In addition to these papers we also invited experts from industry, technology providers and implementers of novel technology to present challenges, real problems and tools in search and search-related areas, delivered as talks in the Industry and Demo Sessions. We would like to thank all of them for presenting novel and innovative search tools and UIs and sharing with researchers other ideas related to the process of information finding.

Our sincere thanks go out to: the Local Organizing Committee, Andreas Rauber, Allan Hanbury, Mihai Lupu, Marie-Pierre Garnier, Christiane Tronigger and Christine Haas; Norbert Fuhr for his support as General Chair of the conference; the IRF scientific board for agreeing that we organize the conference; the keynote speaker; the members of the Program Committee and the additional reviewers for their thorough reviews; the Municipality of Vienna for their support and for providing us with the City Hall to host the conference dinner; the MUMIA COST Action, the Vienna University of Technology, and the Austrian Computer Society for their organizational assistance, and, to Microsoft Research for providing sponsorship to the conference.

We hope that you enjoy the conference proceedings.

April 2012   Michail Salampasis
                                                                         Birger Larsen

Organization

Program Committee

Galia Angelova	Institute for Parallel Processing, Bulgarian Academy of Sciences, Bulgaria
Pavel Braslavski	Yandex
Paul Buitelaar	DERI - National University of Ireland, Galway, Ireland
Pablo Castells	Universidad Autónoma de Madrid, Spain
Ivan Chorbev	FEIT
Bruce Croft	University of Massachusetts Amherst, USA
Danco Davcev	UIST Ohrid
Michael Dittenbach	Max Recall
Gideon Dror	The Academic College of Tel-Aviv-Yaffo, Israel
Anna Esposito	Second University of Naples, Italy
Sebastien Ferre	Université de Rennes 1, France
Wilfried Gansterer	University of Vienna, Austria
Gregory Grefenstette	Exalead
Allan Hanbury	Vienna University of Technology, Austria
David Hawking	Funnelback
Ivan Koichev	University of Sofia "St. Kliment Ohridski", Bulgaria
Yiannis Kompatsiaris	CERTH - ITI
Udo Kruschwitz	University of Essex, UK
David Lamas	Tallinn University, Estonia
Birger Larsen	Royal School of Library and Information Science
Christina Lioma	University of Copenhagen, Denmark
David Losada	University of Santiago de Compostela, Spain
Mihai Lupu	Vienna University of Technology, Austria
Bente Maegaard	University of Copenhagen, Denmark
Walid Magdy	Dublin City University, Ireland
Marie-Francine Moens	Katholieke Universiteit Leuven, Belgium
Igor Mozetic	Jozef Stefan Institute, Slovenia
Henning Müller	HES-SO
Hidetsugu Nanba	Hiroshima City University, Japan
Ragnar Nordlie	Oslo University College, Norway
Andreas Nuernberger	University of Magdeburg, Germany
Georgios Paltoglou	University of Wolverhampton, UK
Andreas Rauber	Vienna University of Technology, Austria

Patrick Ruch	University Hospital of Geneva, Switzerland
Tony Russell-Rose	UXLabs
Ian Ruthven	University of Strathclyde, UK
Michail Salampasis	Vienna University of Technology, Austria
Jan Snajder	University of Zagreb, FER, Croatia
Oren Somekh	Yahoo! Labs
Hannu Toivonen	University of Helsinki, Finland
Elaine Toms	University of Sheffield, UK
Dolf Trieschnigg	University of Twente, The Netherlands
Howard Turtle	Syracuse University, USA
Ioannis Tzitzikas	ICS-FORTH, University of Crete, Greece
David Vallet	Universidad Autónoma de Madrid, Spain
Suzan Verberne	Centre of Language and Speech Technology, Radboud University Nijmegen, The Netherlands
Thijs Westerveld	Teezir Search Solutions

Additional Reviewers

Arcan, Mihael
Nitsche, Marcus
Sébillot, Pascale

Table of Contents

Query Formulation and Analysis

Learning-Based Pseudo-Relevance Feedback
for Patent Retrieval

Parvaz Mahdabi and Fabio Crestani

Faculty of Informatics, University of Lugano, Switzerland
{parvaz.mahdabi,fabio.crestani}@usi.ch

Abstract. Pseudo-relevance feedback (PRF) is an effective approach in Information Retrieval but unfortunately many experiments have shown that PRF is ineffective in patent retrieval. This is because the quality of initial results in the patent retrieval is poor and therefore estimating a relevance model via PRF often hurts the retrieval performance due to off-topic terms. We propose a *learning to rank* framework for estimating the effectiveness of a patent document in terms of its performance in PRF. Specifically, the knowledge of effective feedback documents on past queries is used to estimate effective feedback documents for new queries. This is achieved by introducing features correlated with feedback document effectiveness. We use patent-specific contents to define such features. We then apply regression to predict document effectiveness given the proposed features. We evaluated the effectiveness of the proposed method on the patent prior art search collection CLEF-IP 2010. Our experimental results show significantly improved retrieval accuracy over a PRF baseline which expands the query using all top-ranked documents.

Keywords: Patent Retrieval, Pseudo-Relevance Feedback, Query Modeling, Prior-art Search.

1   Introduction

Patent prior art search is composed of a search over previously filed patents with the aim of retrieving relevant documents, which may invalidate or at least describe the prior art work in a patent application, (henceforth referred to as *query patent* in this paper). The challenges of patent prior art search are different from those of standard ad hoc text and web search. The first difference is associated to the query length: patent prior art queries are full patent applications comprising of hundreds of words as opposed to ad hoc search and web search where the queries are very short. The second issue is related to the fact that patent prior art search is a recall oriented task where the goal is to retrieve all relevant documents at early rank positions as opposed to ad hoc and web search, where the goal is to achieve high precision.

Query expansion using pseudo-relevance feedback (PRF) is one of the most effective practices in information retrieval [22] which can improve the average retrieval effectiveness. PRF usually use the top retrieved documents based on

M. Salampasis and B. Larsen (Eds.): IRFC 2012, LNCS 7356, pp. 1–11, 2012.
© Springer-Verlag Berlin Heidelberg 2012

their retrieval scores for the initial query. The basic assumption of PRF is that the top-ranked documents in the initial retrieval run contain many useful terms and thus can help identify the relevant documents from the non-relevant ones. In general, the expansion terms are selected either according to the term distributions in the feedback documents or according to the comparison between the term distributions in the feedback documents and in the collection.

However, since not all the top retrieved documents are relevant, there is a high chance that we use non-relevant documents for expansion. Few studies [4,2] show that although PRF improves the retrieval effectiveness for many queries, it hurts many other queries and this challenges the usefulness of PRF in real retrieval applications. There are two factors that affect the retrieval performance of PRF: i) the quality of the feedback documents used for the expansion phase; and ii) the expansion terms extracted from the set of feedback documents. Recently, machine learning was used to improve the PRF's effectiveness from both of these aspects.

Cao et al. proposed to refine PRF at a term level [1]. They use a Support Vector Machine (SVM) to select good expansion terms using a set of term-level features such as the proximity of an expansion term and an original query term, or the co-occurrences of an expansion term and an original query term in the collection. He and Ounis proposed to improve PRF by choosing the right documents for relevance feedback [8]. They apply Naïve Bayes classification and Logistic Regression to classify feedback documents. They use document-level features such as the distribution of query terms in the feedback document and all the top-ranked documents or the proximity between the expansion terms and the original query terms in the feedback document. They then only use the documents predicted to be of good quality for query expansion. Recently, Keikha et. al [11] proposed a regression model for predicting the effectiveness of feedback documents. They use various document-level features such as the similarity of feedback documents with other top retrieved documents or with the collection. All these works achieve effective retrieval performance by selecting the most appropriate feedback terms or feedback documents.

The use of PRF on news and web data has been shown to improve the retrieval effectiveness. Unlike other fields, existing studies show that PRF does not perform well on patent documents because of the original poor quality of the results [21,16]. In this work we try to address this problem by selecting the most appropriate feedback documents to be used in the expansion phase.

In this paper, we present a novel method to incorporate relevance feedback into a learning to rank framework by exploiting patent-specific contents.

Our contributions are:

- Presenting a method for predicting the effectiveness of a feedback document using regression.
- Introducing different features to capture the effectiveness of a feedback document in terms of its performance in query expansion.

We evaluate our model on the CLEF-IP 2010 collection and we report significant improvement over a strong PRF baseline. The results show that by incorporating the predicted effectiveness of a feedback document, we can achieve significant improvement over using all top-ranked documents for expansion.

The rest of this paper is organized as follows: Section 2 briefly reviews the related work; Section 3 defines the original and the expanded query models; Section 4 describes the prediction model using the proposed features; Section 5 reports the experimental results and Section 6 reports the conclusions of the work.

2   Related Work

The work performed by patent examiners involves manual query formulation from the query patent in order to find invalidating claims. They consider high term frequency in the document to be a strong indicator of a good query term. Methods to shorten a query patent have been studied for few years and this research direction has shown to be very challenging mainly due to the presence of ambiguous terms. In the third NTCIR workshop [10], the first patent prior art search track was introduced and several patent test collections were released. Some early works [9,12] using this collection focused on extracting keywords to form a reduced query.

A recent line of work advocates the use of the full patent application as the query to reduce the burden on patent examiners. This direction has been started by Xue and Croft [23], who conducted a series of experiments in order to examine the effect of different patent fields, and concluded with the observation that the best Mean Average Precision (MAP) is achieved using the text from the *description* section of the query patent with raw term frequencies. Fuji [6] showed that retrieval effectiveness can be improved by combining IR methods with the result of citation extraction.

The current developments in the patent search are driven by the Intellectual Property task within the CLEF[1] initiative. Several teams participated in the prior art search task of the CLEF-IP 2010 and proposed approaches to reduce the query patent by extracting a set of key terms from it. Different participating teams experimented with term distribution analysis in a language modeling setting, and they employed the document structure of the patent documents in various ways [19].

So far, one of the most comprehensive descriptions of the problem and possible solutions for the prior art search is presented by Magdy and Lopez [17]. Authors show that the best performing run of CLEF-IP 2010 uses citations extracted by training a Conditional Random Field (CRF). The second best run uses a list of citations extracted from the patent numbers within the *description* field of some patent queries. They also show that the best run employs sophisticated methods of retrieval using two complementary indices, one constructed by extracting terms from the patent collection and the other built from external resources such as Wikipedia. They compare these two approaches and conclude with an interesting observation that the second best run achieves a statistically

---

[1] http://www.ir-facility.org/clef-ip

indistinguishable performance compared to the best run. As a baseline for this paper, we consider an approach which produces comparable results to the second rank participating group of CLEF-IP 2010 and we compare our proposed method to this baseline.

A recent work [7] studies the effect of using PRF for reducing patent queries. Authors decompose a patent application into constituent text segments and compute the language modeling similarities by calculating the probability of generating each segment from the top ranked documents. This work achieves improvement over initial results using PRF which is satisfactory compared to all standard PRF methods which failed in the patent domain. However, their approach is not able to achieve statistical significance gain over the second best result of CLEF-IP 2010. We also compare the performance of our proposed method with this work.

In this paper, in addition to the well known MAP metric, we use the Patent Retrieval Evaluation Score (PRES) which is originally proposed by Magdy and Jones [15]. Authors showed that MAP can be a misleading metric for evaluating the performance of patent prior art search because of its inherent characteristic of favoring precision over recall. This metric measures the system recall and the quality of ranking in one single score.

3   Establishing a Baseline

Patent prior art queries are full patent applications which are much longer than standard web search topics. These queries, being composed of hundreds of terms, do not represent a focused information need. Thus, the success of the patent prior art search relies on the selection of good search queries.

Our goal is to estimate the query model of a query patent in a language modeling framework. This transformation enables us to identify the importance of terms and assign weights to them accordingly. By modeling the term distribution of the query patent we get a detailed representation of the query patent which allows us to expand the query, and to refine the query model by considering relationships between terms. This approach is used to bridge the vocabulary gap between the underlying information need of the query patent and the collection.

3.1   Estimating the Query Model

In this section, we first describe how we create a language model θ_Q for the query patent. We use the maximum likelihood estimate smoothed by the background language model, as expressed in Equation 1 to avoid sparseness issues.

$$P(t|\theta_Q) = (1 - \lambda) \cdot P_{ML}(t|D) + \lambda \cdot P_{ML}(t|C) \qquad (1)$$

where maximum likelihood estimate P_{ML} is calculated as follows:

$$P_{ML}(t|D) = \frac{n(t, D)}{\sum_{t'} n(t', D)}$$

We introduce a unigram query model by estimating the importance of each term according to a weighted log-likelihood based approach as expressed below:

$$P(t|\theta_{orig}) = Z_t \, P(t|\theta_Q) \, log\left(\frac{P(t|\theta_Q)}{P(t|\theta_C)}\right)$$

where $Z_t = \frac{1}{\sum_{t \in V} P(t|\theta_{orig})}$ is the normalization factor and is defined as the Kullback-Leibler divergence between θ_Q and θ_C. This approach favors terms that have high similarity to the document language model θ_Q and low similarity to the collection language model θ_C [18]. We refer to this original query model as *LM*.

3.2 Estimating the Relevance Model

The relevance model is used to select the most appropriate terms (feedback terms) from the top retrieved documents by the initial search (feedback documents) to expand the original query [13]. Feedback terms for query expansion are selected based on their relevance to the initial query:

$$P(w|\theta_R) \propto \sum_{D \in \mathbb{D}} P(w|D)P(Q|D) \tag{2}$$

where $\mathbb{D}$ is a set of feedback documents for the initial query Q.

In a PRF scenario, an underlying relevance model θ_R is inferred by combining the language models of the top retrieved documents where the weight of each document is assumed to be proportional to its score for the initial query.

In our experiments, the estimated relevance model is interpolated with the original query model as follows:

$$P(w|\theta_{exp}) = (1 - \lambda) \, P(w|\theta_{orig}) + \lambda \, P(w|\theta_R) \tag{3}$$

where λ is the interpolation weight that is set to 0.5 in our experiments. We refer to this expanded query model using relevance model as *RM*. We use the query likelihood as the weight of each feedback document.

4   Finding Effective Feedback Documents

Our goal is to identify useful feedback documents for query expansion. To this end we extract a set of features that can help us predict the effectiveness of each document. These features are mainly based on the content of each document and its relation with the other top retrieved documents and the collection. We apply some of the features used in [8,11] in our work. In order to capture the specific nature of the patent documents we use patent-specific characteristics for defining the features. These features are described in this section.

- *Relevance Score*: this feature is an indicator of a good feedback document in the standard PRF. This relevance score assumes that the higher the document is ranked in the first retrieval pass, the higher the chance that it can be a good feedback document. We use the relevance score generated by three ranking models, namely Okapi BM25 [20], Language model using Dirichlet smoothing [24] and Positional Language Model [14] of each feedback document. We used Gaussian density kernel and we set the Dirichlet prior to 2500 in our experiments.
- $\delta(\mathbb{D}, D)$: this feature measures the similarity between a feedback document D and the whole feedback document set $\mathbb{D}$. We define a distance function based on the Kullback-Leibler (KL) divergence. To compensate for the terms which are absent in each of these distributions, we consider the symmetric version of the KL-divergence between the two distributions $\mathbb{D}$ and D.

$$\delta(\mathbb{D}, D) = \frac{1}{2}[D_{KL}(D||\mathbb{D}) + D_{KL}(\mathbb{D}||D)]$$

- *Clarity*: This measure quantifies the level of effectiveness of a query at retrieving a specific topic [3]. The clarity measure is the KL-divergence between the feedback language model $P(w|D)$ and the collection language model $P(w|C)$.

$$D_{KL}(D||C) = \sum_{w \in V} P(w|D) \cdot log\frac{P(w|D)}{P(w|C)}$$

- *Topical Clarity*: this measure is defined as the KL-divergence between the language model of feedback document D and the language model of IPC_Q. We propose this measure inspired by the clarity measure using patent-specific characteristics. Let IPC_Q be the set of documents dealing with similar topics to Q, as represented by IPC classes[2]. Formally, the topical clarity measure is defined as:

$$D_{KL}(D||IPC_Q) = \sum_{w \in V} P(w|D) \cdot log\frac{P(w|D)}{P(w|IPC_Q)}$$

where $P(w|IPC_Q)$ is the relative frequency of term w in documents with similar IPC classes to Q. We refer to this as the topical clarity. In this case, a larger KL-divergence indicates a query with fewer topics and therefore a more focused query, while a smaller KL-divergence indicates a query with a broader language use.
- *IPC-based Clarity*: this measure captures the similarity between the language usage of IPC_D and the collection language model $P(w|C)$. Let IPC_D be the set of documents dealing with similar topics to feedback document D, as represented by IPC classes. This measure is defined as:

$$D_{KL}(IPC_D||C) = \sum_{w \in V} P(w|IPC_D) \cdot log\frac{P(w|IPC_D)}{P(w|C)}$$

---

[2] http://www.wipo.int/classifications/ipc/en/

To learn a model for predicting the effectiveness of a feedback document using the above features we define the following regression problem.

$$arg \min_{\Phi} \sum_{Q \in T} \sum_{D \in \mathbb{D}} ||\Phi(F(D)) - AP(D)||^2 \tag{4}$$

where T is a set of training topics, $\mathbb{D}$ is a set of the 100 top retrieved documents for query Q, and F is a mapping from feedback document D to the feature space. We predict the Average Precision (AP) for each feedback document according to Equation 4 and we use this value as its effectiveness. This value is used as a weight which shows the contribution of a feedback document to the estimation of the relevance model.

In order to model the relationships between combinations of predictor variables and AP, we use a Stochastic Gradient Boosting Tree (SGBT) [5]. This model produces an ensemble of weak prediction learners, i.e., decision trees. It builds additive regression models in a stage-wise manner and it generalizes them by allowing optimization of an arbitrary differentiable loss function. For the SGBT, we used the gbm2 package implemented in R[3]. SGBT can find a sub-combination of features that may aid with the prediction of AP of a feedback document.

5   Experimental Methodology and Results

In this section, we present the results for an experimental evaluation of query expansion on patent documents using feedback effectiveness predictors. First, we describe our experimental setup in this study. We then investigate the possible impact of selecting a good feedback document on query expansion. We then conduct an experiment to estimate the effectiveness of a feedback document based on the set of features proposed in Section 4.

5.1   Experimental Setup

The retrieval experiments described in this paper are implemented using Terrier[4]. We used the CLEF-IP 2010 collection as or dataset which consists of 2.6 million patent documents. The relevance assessments are provided for the topic set which are patent applications that have *title*, *abstract*, *description*, and *claims* sections. As mentioned earlier using description text for query generation has been shown to achieve the best MAP in contrast to other patent fields [23]. Therefore, we used the description text for building the query model. Patent applications in the topic set are annotated with the metadata tags such as IPC classes. We worked with the English topic set which corresponds to 1348 topics. During indexing and retrieval, both documents and queries were stemmed using the Porter stemmer. Stop-word removal was performed on both documents

---

[3] http://cran.r-project.org/web/packages/gbm/
[4] http://terrier.org/

and queries using the standard Terrier stop-word list. In addition to that, we also removed all the formulas and numeric references. We used a five-fold cross validation for our experiments.

5.2   Usefulness of Effective Feedback Documents

In this section we investigate the possible impact of selecting a good feedback document on query expansion. To this end, we use the true value of AP score of each expanded query as the effectiveness value of the feedback document used for generating the expanded query. The effectiveness value of a feedback document is used as the weight of the feedback documents (instead of query likelihood) in the estimation of relevance model as expressed in Equation 2.

We aggregate the document models of feedback documents by the weighted average and generate a single relevance model. We then interpolate the relevance model with the original query following Equation 3. We use the top 10 documents with the highest effectiveness values to estimate the relevance model. We refer to this model as $oracle_{exp}$ which shows the potential upper bound of what can be achieved by using the true effectiveness values.

Table 1. Performance results using the true feedback document effectiveness

model	MAP
LM	0.136
RM	0.103
$oracle_{exp}$	0.195 † ‡

Table 1 shows the MAP for the top 1000 results with expansion method $oracle_{exp}$ in comparison with the original unigram query model (LM) and the expanded query model (RM) using the relevance model.

As the results of Table 1 suggests, $oracle_{exp}$ achieves 89% improvement in terms of MAP over RM and 43% over LM, by employing the true effectiveness value of a feedback document. This shows a great room for improvement over standard PRF and we seek to reach this upper bound by a reasonable estimation of the correct AP values. Results marked with † and ‡ indicate that the achieved improvement of $oracle_{exp}$, over LM and RM, is statistically significant at $p < 0.05$ using t-test.

5.3   Predicting the Effectiveness of a Feedback Document

In order to find out whether the defined features can indicate a successful estimation of relevance model for query expansion, we predict the effectiveness of a feedback document. Table 2 presents the results of query expansion on patent documents using feedback effectiveness predictors ($Pred$-PRF). We compare our results with the original query model (LM) and the expanded query model using the relevance model (RM). We also compare our work with the query reduction (PQR) method of Ganguly et al. [7]. Finally, we compare our results with the oracle expansion $oracle_{exp}$.

Table 2. Retrieval results using all methods on the CLEF-IP 2010 collection

model	MAP	PRES
LM	0.136 †	0.535 †
RM	0.103	0.481
PQR	0.137	0.467
$Pred$-PRF	0.158 † ‡	0.558 † ‡
$oracle_{exp}$	0.195	0.603

As the results of Table 2 suggests standard PRF (RM) fails to improve over LM. This is due to the fact that the relevance model RM usually uses the top retrieved documents based on their initial retrieval scores. However, since not all the top retrieved documents are relevant, there is a high chance that we use non-relevant documents for expansion. This problem is exacerbated in patent domain because of the poor quality of the initial results.

Our query expansion using the prediction model ($Pred$-PRF) achieves highly signifiant MAP improvement over both LM and RM on the test collection of CLEF-IP 2010. $Pred$-PRF also achieves 15% improvement in terms of MAP compared to PQR. Results marked with † and ‡ indicate the significant improvement over RM and LM using t-test at $p < 0.05$. These improvements hold for PRES as well. The results of Table 2 indicate that the chosen features were able to predict the effectiveness of a feedback document for estimating the relevance model. This also suggests that the predicted AP was a good indicator of the effectiveness of a feedback document. However, the prediction method did not achieve the upper bound performance $oracle_{exp}$, which is due to the error made by the prediction model. We need to define better features to reach this upper bound. We note that the result of Table 2 is reported using 10 feedback documents and 30 terms.

We calculated the influential features from the learnt SGBT model [5]. The relevance scores calculated by Positional Language Model and BM25 along with the topical clarity and IPC-based clarity are the most influential features.

5.4 Parameter Sensitivity

In this section, we conduct experiments to study the impact of using different number of feedback documents on the retrieval effectiveness of our proposed method $Pred$-PRF. The results are shown in Figure 1. As the Figure 1 suggests RM shows a high sensitivity against the number of feedback documents but the results of $Pred$-PRF are more stable against different sizes of the pseudo-relevance set. In fact, this indicates that our proposed method is able to select the useful feedback documents for different sizes of the pseudo-relevant set.

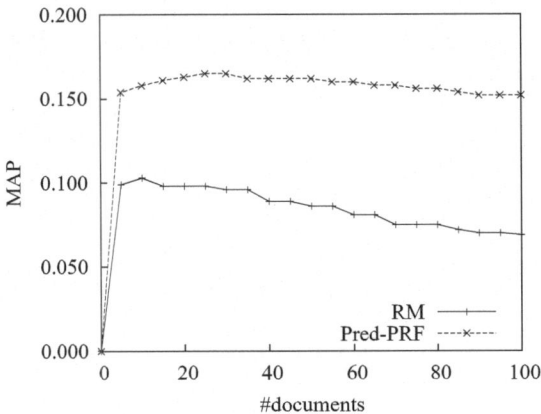

Fig. 1. Sensitivity of RM and Pred-PRF against the number of feedback documents used for relevance feedback

6   Conclusions

This paper presented a learning to rank framework for predicting the effectiveness of a feedback document for query expansion. We proposed a set of features correlated with the effectiveness of a feedback document by utilizing patent-specific characteristics. We then applied regression to predict the effectiveness as a function of the introduced features. We experimentally determined the upper bound of what can be achieved by looking into the true effectiveness of feedback documents. Our prediction method using the proposed features obtained a statistically significant improvement over standard pseudo-relevance feedback. One way to extend our work is to select effective expansion terms for improving pseudo-relevance feedback. This approach is complementary to our current work which only selects effective feedback documents. A possible future work is to introduce a framework that combines these two strategies.

Acknowledgements. Authors would like to thank Information Retrieval Facility (IRF) for the support of this work.

References

1. Cao, G., Nie, J.-Y., Gao, J., Robertson, S.: Selecting good expansion terms for pseudo-relevance feedback. In: Proceedings of SIGIR, pp. 243–250 (2008)
2. Collins-Thompson, K.: Reducing the risk of query expansion via robust constrained optimization. In: Proceedings of CIKM, pp. 837–846 (2009)
3. Cronen-Townsend, S., Zhou, Y., Croft, W.B.: Predicting query performance. In: Proceedings of SIGIR, pp. 299–306 (2002)
4. Dillon, J.V., Collins-Thompson, K.: A unified optimization framework for robust pseudo-relevance feedback algorithms. In: Proceedings of CIKM, pp. 1069–1078 (2010)

5. Friedman, J.H.: Stochastic gradient boosting. Computational Statistics and Data Analysis 38, 367–378 (1999)
6. Fujii, A.: Enhancing patent retrieval by citation analysis. In: Proceedings of SIGIR, pp. 793–794 (2007)
7. Ganguly, D., Leveling, J., Magdy, W., Jones, G.J.F.: Patent query reduction based on pseudo-relevant documents. In: Proceedings of CIKM, pp. 1953–1956 (2011)
8. He, B., Ounis, I.: Finding good feedback documents. In: Proceedings of CIKM, pp. 2011–2014 (2009)
9. Itoh, H., Mano, H., Ogawa, Y.: Term distillation in patent retrieval. In: Proceedings of the ACL 2003 Workshop on Patent Corpus Processing, pp. 41–45 (2003)
10. Iwayama, M., Fujii, A., Kando, N., Takano, A.: Overview of the third NTCIR workshop. In: Proceedings of the ACL 2003 Workshop on Patent Corpus Processing, pp. 24–32 (2003)
11. Keikha, M., Seo, J., Croft, W.B., Crestani, F.: Predicting document effectiveness in pseudo relevance feedback. In: Proceedings of CIKM, pp. 2061–2064 (2011)
12. Konishi, K.: Query terms extraction from patent document for invalidity search. In: Proceedings of NTCIR 2005 (2005)
13. Lavrenko, V., Croft, W.B.: Relevance-based language models. In: Proceedings of SIGIR, pp. 120–127 (2001)
14. Lv, Y., Zhai, C.: Positional language models for information retrieval. In: Proceedings of SIGIR, pp. 299–306 (2009)
15. Magdy, W., Jones, G.J.F.: PRES: A score metric for evaluating recall-oriented information retrieval applications. In: Proceedings of SIGIR, pp. 611–618 (2010)
16. Magdy, W., Leveling, J., Jones, G.J.F.: Exploring Structured Documents and Query Formulation Techniques for Patent Retrieval. In: Peters, C., Di Nunzio, G.M., Kurimo, M., Mostefa, D., Penas, A., Roda, G. (eds.) CLEF 2009. LNCS, vol. 6241, pp. 410–417. Springer, Heidelberg (2010)
17. Magdy, W., Lopez, P., Jones, G.J.F.: Simple vs. sophisticated approaches for patent prior-art search. In: Proceedings of ECIR, pp. 725–728 (2010)
18. Mahdabi, P., Keikha, M., Gerani, S., Landoni, M., Crestani, F.: Building Queries for Prior-Art Search. In: Hanbury, A., Rauber, A., de Vries, A.P. (eds.) IRFC 2011. LNCS, vol. 6653, pp. 3–15. Springer, Heidelberg (2011)
19. Piroi, F., Tait, J.: CLEF-IP 2010: Retrieval experiments in the intellectual property domain. In: Workshop of the Cross-Language Evaluation Forum, LABs and Workshops, Notebook Papers, CLEF 2010 (2010)
20. Robertson, S.E., Walker, S.: Some simple effective approximations to the 2-poisson model for probabilistic weighted retrieval. In: Proceedings of SIGIR, pp. 232–241 (1994)
21. Takeuchi, H., Uramoto, N., Takeda, K.: Experiments on Patent Retrieval at NTCIR-5 Workshop (2005)
22. Xu, J., Croft, B.: Query expansion using local and global document analysis. In: Proceedings of SIGIR, pp. 4–11 (1996)
23. Xue, X., Croft, W.B.: Transforming patents into prior-art queries. In: Proceedings of SIGIR, pp. 808–809 (2009)
24. Zhai, C., Lafferty, J.: A study of smoothing methods for language models applied to ad hoc information retrieval. In: Proceedings of SIGIR, pp. 334–342 (2001)

Analyzing Parallelism and Domain Similarities in the MAREC Patent Corpus

Katharina Wäschle and Stefan Riezler

Department of Computational Linguistics, Heidelberg University
{waeschle,riezler}@cl.uni-heidelberg.de

Abstract. Statistical machine translation of patents requires large a-mounts of sentence-parallel data. Translations of patent text often exist for parts of the patent document, namely title, abstract and claims. How-ever, there are no direct translations of the largest part of the document, the description or background of the invention. We document a twofold approach for extracting parallel data from all patent document sections from a large multilingual patent corpus. Since language and style differ depending on document section (title, abstract, description, claims) and patent topic (according to the International Patent Classification), we sort the processed data into subdomains in order to enable its use in domain-oriented translation, e.g. when applying multi-task learning. We investigate several similarity metrics and apply them to the domains of patent topic and patent document sections. Product of our research is a corpus of 23 million parallel German-English sentences extracted from the MAREC patent corpus and a descriptive analysis of its subdomains.

1 Introduction

Statistical machine translation (SMT) requires large amounts of parallel data on the sentence level to train translation and language models of high coverage. Best results are obtained if parallel data are available for the specific domain in question. Patent translation is particularily dependent on the availability of large in-domain parallel data sets for several reasons: Patent translation deals with documents that exhibit a highly specialized vocabulary, consisting of tech-nical terms specific to the field of invention the patent relates to and legal jargon ("patentese") that is not found in everyday language. To maximize their cover-age, patents are often intentionally vague and ambiguous. Furthermore, patents exhibit a complex textual structure of differently designated text fields. Some patent documents contain translations; however, in most cases only parts of the patent, namely titles and abstracts or claims, are multilingual, while for the largest document section, the description, no direct translation is available. This poses a potential data sparsity problem for patent translation.

In this paper we investigate possibilities for building resources for patent translation by extracting large amounts of parallel data from a multilingual patent document corpus, MAREC, and preparing the data for domain-oriented translation. Multi-task learning, for example, aims to enhance machine learning

M. Salampasis and B. Larsen (Eds.): IRFC 2012, LNCS 7356, pp. 12–27, 2012.

performance by learning tasks on several separate but similar domains at the same time. Patents differ with respect to vocabulary – e.g. patents assigned to IPC section C (chemistry) tend to contain a large amount of chemical formula – and style – the patent title consists of a single noun phrase, while claims exhibit a complex phrasal structure. This induces different subdomains that can be viewed as tasks in multi-task translation. First experiments on multi-task patent translation on tasks defined by patent topic and document sections have been presented by [1].

The focus of this paper lies on the corpus construction and description, so we employ patent translation as a tool for similarity analysis. We train separate translation models on every subdomain and evaluate across domains in order to investigate similarities and differences between domains. Furthermore, we apply several information-theoretic similarity metrics to the topic dimension of IPC patent classification. While in general, every subdomain is best translated with a model trained on the respective domain, we find a correlation between similarity of IPC domains as measured by information-theoretic metrics and BLEU evaluation in cross-domain translation over IPC domains. This shows that information-theoretic measures can be used to select appropriate patent texts from related domains for augmenting training data. Furthermore, we show that due to particularities of the patent data language-specific preprocessing, such as German compound splitting, can be a key technique for German-English patent translation, especially for the translation of titles.

Product of our research is a corpus of over 23 million parallel German-English sentences from all IPC domains and text sections, sorted accordingly. Together with the descriptive analysis given in this paper, this forms an enabling resource for research on patent translation and tasks that build on translation, such as cross-lingual patent retrieval.

2   Related Work

Patent translation is an active research area that is fueled by benchmark testing workshops such as CLEF[1] and NTCIR[2]. NTCIR features a patent translation task for Japanese-English and Chinese-English patent documents. The data for the former task contains about 2 million sentence pairs that were automatically extracted from the description sections. The extraction method is described in [2]. It involves a pipeline architecture where in a first step length-based alignment scores ([3]) are used to propose sentence alignment candidates, which are then filtered using dictionary-based word translation scores. [2] also report results on patent translation experiments across IPC sections, showing that MT results are best when training and test sections coincide. Furthermore, pooling data from all sections for a maximum amount of training data achieved best results overall. [4] extract 160,000 Chinese-English sentence pairs using a pipeline of candidate

---

[1] Cross-Language Experiment Forum, http://www.clef-campaign.org
[2] National Institute for Informatics Test Collection for IR Systems, http://research.nii.ac.jp/ntcir/

sentence alignments that are filtered based on lexical translation scores. They do not report cross-section MT experiments.

The MAREC dataset is a superset of the patent retrieval data used in the CLEF-IP tracks, and has been deployed for stand-alone patent translation by [5], [6], and [7]. Again, an extraction procedure combining a candidate sentence alignment with a word-based translation filter is used. For example, [6] combines Gargantua ([8]) for sentence alignment with GIZA++ ([9]) for word alignment. Cross-section MT experiments for IPC domains are reported in [5] and [7], where the maximum-sized pool of combined data from all sections yields the best results, especially for language modeling. The sentence-parallel data extracted from MAREC for the experiments in [5], [6], and [7] is not publicly available. Furthermore, only data from abstracts and claims sections was extracted.

3   Structural and Topical Dimensions in Patent Text

We analyze patents with respect to the domain dimensions of both topic – the technical field covered by the patent – and structure – a patent's text sections – with respect to their influence on machine translation performance.

The topic dimension of patents is given by the International Patent Classification (IPC)[3] which categorizes patents hierarchically into 8 sections, 120 classes, 600 subclasses, down to 70,000 subgroups at the leaf level. Table 1 shows the 8 top level sections. A patent can be assigned to multiple IPC classes.

Table 1. IPC top level sections

A Human Necessities
B Performing Operations, Transporting
C Chemistry, Metallurgy
D Textiles, Paper
E Fixed Constructions
F Mechanical Engineering, Lighting, Heating, Weapons
G Physics
H Electricity

In addition to the patent classification, we argue that patents can be subcategorized along the dimension of textual structure. Exemplary, the European Patent Convention (EPC) lays out the structure of a patent[4] in Article 78.1:

"A European patent application shall contain:
(a) a **request for the grant** of a European patent;
(b) a **description** of the invention;

---

[3] http://www.wipo.int/classifications/ipc/en/
[4] Highlights by the authors.

(c) one or more **claims**;

(d) any drawings referred to in the description or the claims;

(e) an **abstract**,

and satisfy the requirements laid down in the Implementing Regulations."

The textual elements of a patent are the title, which is specified in the request for grant, description, claims, and abstract. Examples for each text type can be found in table 2. The title is a short descriptive noun phrase, while the claim exhibits a particular sentence structure.

Table 2. Sample sentences from patent text sections

title	Contact lense forming machine
abstract	Parameters for mold materials and important dimensions are also disclosed.
description	FIGS. 7 and 8 illustrate the final curvatures of a finished plus and minus lense, respectively.
claim	The machine as set forth in claim 2, wherein said lense holding element is secured to the fixed support and the cooperating tool element is mounted on the first pivotal support means.

4 Extraction of Parallel Text

Our work is based on the MAREC[5] patent data corpus. It contains over 19 million patent applications and granted patents from four patent organizations (European Patent Office (EP), World Intellectual Property Organization (WO), United States Patent and Trademark Office (US), Japan Patent Office (JP)), from 1976 to 2008 in a standardized format. We extract data for our experiments from the EP and WO subcorpora which contain multilingual patent documents that feature partial translations of the patent text between German, English and French (the EPO's official languages). We assume translated titles to be sentence-aligned by default, and define multilingual document sections, which are of similar length in both languages as parallel[6]. To extract parallel text sections, we first determine the longest instance of the respective section, if different document kinds[7] exist for a patent. Overall, we extracted 2,204,384

---

[5] http://www.ir-facility.org/prototypes/marec

[6] We compute the number of German tokens relative to the number of English tokens and keep parallel sections with a ratio larger than 0.7.

[7] A patent kind code indicates the document stage in the filing process, e.g., A for applications and B for granted patents, with publication levels from 1-9. See http://www.wipo.int/standards/en/part_03.html

parallel titles, 291,716 parallel abstracts, and 735,667 parallel claims sections for the German-English language pair. However, there are no parallel descriptions.

The lack of directly translated descriptions poses a serious limitation for patent translation, since this section constitutes the largest part of the document. It has been shown that it is possible to obtain comparable descriptions from related patents that have been filed in different countries and are connected through the patent family id. [2] introduce this method to collect Japanese-English patent translations, [4] apply the same technique to Chinese and English patents. We transfer this approach to German-English patents and search the US collection of MAREC for documents sharing a family id with EP patents that feature a German description. We extracted 172,472 patents that were both filed with the USPTO and the EPO and contain an English and a German description, respectively. However, data extracted in this way is presumably less parallel than the directly translated sections. This is due to the different origination process: translations of document sections are drawn up during the application process and aim to stay close to the original text. The filing of an application for the same or a closely related invention in a different country might occur several years after the first publication, during which the invention might have been subject to changes. Due to differing regulations and application procedures there might be further modifications, amendments and omissions compared to the original document. [2] and [4] use a scoring function to determine good translations after performing sentence alignment, deploying an aggregate score of sentence length ratio, IBM Model 1 word translation probabilites and a lexical score using a dictionary. Since we do not have access to a domain-specific dictionary[8], we enforce the first two criteria for the selection of parallel sentences from descriptions in the sentence alignment process.

For alignment, we used Gargantua[9] ([8]), an unsupervised, language pair independent open source aligner, which implements the idea presented in [11]: a combination of both sentence-length-based alignment and a lexical translation model in a two-pass approach. The best alignment is calculated based on sentence length in the first pass, and recalculated with lexical word translation probabilities from an IBM Model 1 estimated on the preliminary alignments in the second pass. The algorithm is robust and can deal with asymmetrical translations, generating one-to-many/many-to-one and 1-to-0/0-to-1 alignments.

Before running the sentence aligner, we cleaned the extracted data and removed several forms of noise. Among the problems we observed were misspellings, hyphens left from line breaks, e.g. Wasser-\nverteilungsnetz, formatting tags such as <IMAGE> or <SEP>, and multiple or missing whitespace.

We eliminated formatting-related problems, such as multiple whitespace, line breaks, tags and hyphens. We did not attempt to correct misspellings, assuming that they will not hurt translation performance significantly. Another problem we observed with MAREC documents are mislabeled language attributes, e.g. a

---

[8] A patent domain dictionary might be created from high-quality multilingual data, for example titles.

[9] http://gargantua.sourceforge.net/

French abstract with English language label EN. We did not perform language detection and therefore cannot quantify the mislabelings, but they appear to be very infrequent; we rely on the sentence aligner, which uses lexical word probabilities, to filter out sentences in the mistaken language.

Sentence aligning requires further preprocessing of the input text, namely splitting the text into sentences and tokenization. We use language-specific, heuristic-based tools distributed with Europarl[10] ([10]), which resolve punctuation ambiguities[11] by using a list of known abbreviations and heuristics for each language, which we extended to include patent-specific abbreviations.

Table 3. Alignment statistics

	de			en	
	output	input	percentage	input	percentage
abstract	720,571	780,161	92.36%	938,117	76.81%
claims	8,346,863	8,533,190	97.82%	8,679,288	96.17%
description	14,082,381	16,330,817	86.23%	17,034,777	82.67%

We aligned the extracted data in batches split by section type and year of origin, in order to speed up the process by parallelisation. Each extracted text section was considered to be a document. Table 3 shows the number of sentences on the source (de) and target (en) side after sentence splitting, compared to the number of aligned sentences that were output by Gargantua. The text from the abstract sections exhibits a strong asymmetry, with about 780,000 sentences on the German corresponding to 940,000 sentences on the English side. Given that extremely unbalanced sections were already discarded in the extraction process, this indicates that abstracts are often not literal translations.This asymmetry also results in a lower input to output ratio of 92.36% in relation to the number of possible alignments, which is given by the number of source sentences. The claims are more balanced with 8.5 million sentences on the source and 8.6 million sentences on the target side and exhibit a high input to output ratio of 97.82%. This is presumably owing to the nature of patent claims: mapped out as a numbered list of sentences, they tend to be translated phrase by phrase. Like the abstracts, the descriptions are quite imbalanced and yield the worst input to output ratio with 86.00% on the source side. Furthermore, spot tests show that the aligned data contains some sentence pairs that only partially overlap.

5   Experimental Data

We conduct information-theoretic as well as MT experiments to analyze and characterize the extracted corpus and gain information that indicates possible

---

[10] http://www.statmt.org/europarl/v6/tools.tgz
[11] A full stop can either indicate the end of a sentence or an abbreviation.

use cases for this data. We split the corpus according to the two previously defined dimensions, namely text sections and top-level IPC classification. In this way, we gain four subcorpora for the structural dimension and eight for the topical, resulting in 24 possible combinations. We furthermore split the data by year of publication date and assign documents published between 1979 and 2007 for training and documents published in 2008 for tuning and testing. To ensure that there is no biasing overlap between training and test data, we compute the percentage of sentences in the extracted data that are actually unique (table 4). Abstract, claims and descriptions are unproblematic with a small amount of duplicates that are partly caused by noise from incorrectly split sentences and partly by sentences that are used in several patents. For instance, the patent documents EP 1050190 (A1), titled Active acoustic devices comprising panel members, and EP 1414266 (A2), titled Active acoustic devices, share four claims word for word. Both patent applications were filed by the same company and name the same inventor. EP 1050190 (A1) was published in 2000, EP 1414266 (A2) four years later. They are obviously related and we can assume that part of the text has been recycled to save writing time, but there is no connection through citation or family id. One could try to prevent such duplicates by exploiting company and inventor metadata, but we argue that these duplicates are a natural characteristic of patent documents and should therefore be kept in the data, if the effect is not biasing evaluation.

For titles, however, the number of duplicates is considerably higher and originates from a different source. Since the title must be short and may only contain technical terms to describe the invention, many patents – from different years and by different companies and inventors – share the same title. For instance, there are 53 patents titled Push button switch in the EP corpus. To prevent a bias, we removed the overlap between test and training set for titles.

Table 4. Percentage of unique sentences relative to the total number of sentences in the experimental data

title	75.09%
abstract	93.92%
description	96.58%
claims	94.65%

We create the actual training, development and test sets by sampling from the subdomains. From every text section subcorpus we sample 500,000 sentences – distributed across all IPC sections – for training and 2,000 sentences each for development and test set. Table 5 gives the number of types and tokens in the resulting training sets. Note that the title set contains considerably fewer tokens than the other sections – only about 15% of the amount contained in the abstracts – so models trained on this section might suffer from the disadvantage of having seen less data than the others. However, the titles still contain a large amount of types – more than half as many as the abstracts – so the disadvantage should not weight too heavily. A potentially hurtful property of the German data is the high type-token-ratio, which is caused by German compounding, especially for titles. We therefore investigate the influence of compound splitting on the translation performance on text sections in an additional experiment.

Table 5. Types and tokens on 500k sentences text section training sets

	de			en		
	#tokens	#types	$\frac{types}{tokens}$	#tokens	#types	$\frac{types}{tokens}$
title	3,267,802	512,773	0.1569	4,038,743	176,293	0.0437
abstract	18,627,983	921,486	0.0495	21,245,542	269,803	0.0127
description	12,836,238	684,190	0.0533	15,961,246	281,053	0.0176
claims	15,646,621	784,978	0.0502	18,355,584	270,013	0.0147

Table 6. Number of sentences per IPC section on claims

A	1,947,542
B	2,522,995
C	2,263,375
D	299,742
E	353,910
F	1,012,808
G	2,066,132
H	1,754,573

For training the IPC domain models, we chose only sentences from the claims and abstract domain, since this data is possibly the cleanest and generally used in IPC cross-domain experiments. We sampled 300,000 sentences from the training corpus for each IPC section. This is the largest training set possible, since the smallest section, D, (see table 6 for the distribution of IPC sections across the claims) contains overall just barely 300,000 sentences in the combined EP and WO training set[12]. [7] use only the five largest sections of the IPC for their experiments, but since we would like to gain a comprehensive view on the data we include all eight IPC domains in our experiments. The resulting number of types and tokens are shown in table 7. There is less variance than on the text sections; still, we can note that section C contains the largest amount of types, which is likely due to the high number of formulae in this section.

---

[12] To address the problem of patent duplicates across different corpora, we only include sentences from the WO which came from documents that do not share a family id with a document in EP.

Table 7. Types and tokens in 300k sentences IPC section training sets

	de			en		
	#tokens	#types	$\frac{types}{tokens}$	#tokens	#types	$\frac{types}{tokens}$
A	9,843,156	520,839	0.0529	11,242,459	233,266	0.0207
B	10,726,633	508,943	0.0474	12,561,139	151,601	0.0121
C	9,514,203	527,609	0.0555	10,942,622	256,932	0.0235
D	13,900,065	440,014	0.0317	16,146,597	160,445	0.0099
E	10,922,892	355,606	0.0326	12,835,170	99,915	0.0078
F	10,941,342	416,113	0.0380	12,941,777	113,498	0.0088
G	10,943,693	578,183	0.0528	12,700,396	180,536	0.0142
H	11,064,367	545,433	0.0493	12,940,731	157,507	0.0122

6   Textual Similarities across IPC Domains

The IPC domains are less well characterized by type-token-ratio, but we expect them to differ with regard to lexical content. To analyize these domains, we therefore compute three information-theoretic similarity measures that perform a pairwise comparison of the vocabulary probability distribution of each task-specific corpus. This distribution is calculated on the basis of the 500 most frequent words in the union of two corpora, normalized by vocabulary size. The first measure is a computation of **Spearman's rank correlation** ([12]) on the frequency-ranked word lists of two corpora. The second measure is a calculation of the **cross-entropy** between the language model probabilities of a model trained on corpus A when applied to corpus B ([13]). As a third metric we use the $\mathcal{A}$-**distance** measure of [14]. If $\mathcal{A}$ is the set of measurable subsets on which the word distributions are defined, then the $\mathcal{A}$-distance is the probability of the subset on which the distributions differ most. A low distance translates to higher similarity.

The three measures for corpus similarity based on the corpus vocabulary are displayed in tables 8, 9 and 10. A low cross-entropy and distance and a high correlation close to 1 or -1 indicate similarity between the vocabulary of two sections. The most similar section or sections – apart from the section itself on the diagonal, highlighted in *italic* font – is indicated in **bold** face. All three measures support a pairwise similarity of A and C, B and F, G and H. Furthermore, a close similarity between E and F is indicated. G and H (electricity and physics, respectively) are very similar to each other but not close to any other section apart from B. This makes sense intuitively, since physics and electricity probably play a more important role in transportations and constructions than, for example, chemistry. Semantically, the two fields are closely related; in fact, the latter can be viewed as a subfield of the former.

The cross-entropy also gives a measure for the homogeneity of each IPC domain on the diagonal; a low perplexity of the language model on a test set from the same domain corresponds to high homogeneity. According to this, C is

Table 8. Pairwise $\mathcal{A}$-distance for 300k IPC training sets

	A	B	C	D	E	F	G	H
A	*0*	0.1303	**0.1317**	0.1311	0.188	0.186	0.164	0.1906
B	**0.1302**	*0*	0.2388	**0.1242**	**0.0974**	**0.0875**	0.1417	0.1514
C	**0.1317**	0.2388	*0*	0.1992	0.311	0.3068	0.2506	0.2825
D	**0.1311**	0.1242	0.1992	*0*	0.1811	0.1808	0.1876	0.201
E	0.188	0.0974	0.311	0.1811	*0*	**0.0921**	0.2058	0.2025
F	0.186	**0.0875**	0.3068	0.1808	**0.0921**	*0*	0.1824	0.1743
G	0.164	0.1417	0.2506	0.1876	0.2056	0.1824	*0*	**0.064**
H	0.1906	0.1514	0.2825	0.201	0.2025	0.1743	**0.064**	*0*

Table 9. Pairwise Spearman's rank correlation for 300k IPC training sets

	A	B	C	D	E	F	G	H
A	*1*	0.5335	**0.5372**	0.5067	0.333	0.3293	0.3192	0.2093
B	**0.5333**	*1*	0.1539	**0.5496**	**0.6132**	**0.6618**	0.4476	0.366
C	**0.5373**	0.1539	*1*	0.3338	-0.0719	-0.0539	0.0756	-0.0226
D	0.5067	0.5496	0.3337	*1*	0.2585	0.2648	0.2636	0.1645
E	0.3329	0.6131	-0.0719	0.2585	*1*	0.6027	0.1928	0.1933
F	0.3293	**0.6618**	-0.0539	0.2648	**0.6027**	*1*	0.2645	0.2684
G	0.319	0.4477	0.0756	0.2636	0.1936	0.2646	*1*	**0.751**
H	0.2091	0.3661	-0.0226	0.1645	0.1933	0.2683	**0.7509**	*1*

the most homogeneous domain, followed by A, which is interesting, since these domains have a comparatively large vocabulary. This may indicate that the cross-entropy is not the best measure to compare different domains.

7   Cross-Domain Translation

We conducted a first investigation of the performance of our corpus in MT. We view these experiments as an expansion of the similarity analysis and therefore only look at cross-domain evaluation. We deliberately do not compare the domain-specific translation models to a model pooled from all data, since previous work has already shown that this outperforms smaller individual models.

We used the phrase-based, open-source SMT toolkit Moses[13] [15] with the standard feature set. We computed 5-gram language models on the target side of the training set with IRSTLM[14] [16] and queried the model with KenLM [17].

The cross-domain evaluation[15] on the IPC classes (table 11) shows that every subdomain is best translated with a model trained on the respective section:

---

[13] http://statmt.org/moses/
[14] http://sourceforge.net/projects/irstlm/
[15] We computed BLEU4 [18] on lowercased data.

Table 10. Cross-entropy: pairwise 300k language model perplexity on 2k IPC test set

	test							
train	A	B	C	D	E	F	G	H
A	*210.8*	320.5	**275.7**	413.4	438.4	417.2	295.6	374.9
B	260.3	*230.0*	332.0	346.5	329.6	**293.5**	**273.3**	**294.6**
C	**220.8**	308.8	*181.0*	**362.6**	490.3	456.6	**281.5**	378.2
D	252.3	248.6	281.1	*239.8*	356.1	332.8	303.3	332.7
E	267.0	254.3	375.9	374.3	*258.0*	**290.3**	299.7	312.9
F	267.1	**239.3**	355.3	**361.0**	**308.6**	*233.8*	**277.9**	**288.1**
G	360.6	367.1	472.9	565.3	539.3	467.5	*255.7*	303.5
H	402.7	386.5	529.5	606.8	546.4	469.5	**280.7**	*257.7*

the BLEU scores on the diagonal are the highest in every column. For assessing similarities, we are therefore interested in the runner-up on each section (indicated in **bold** font). Note that best section scores vary considerably, ranging from 0.5719 on C to 0.4714 on H, indicating classes that are easier to translate. C, the Chemistry section, presumably benefits from the fact that the data contains chemical formulae, which are language-independent and do not have to be translated. [7] show similar variations on the five largest IPC classes with scores ranging from 0.609 on C to 0.5518 on G for the PLuTO data. The higher overall scores and smaller variance are due to the larger amounts of training data used in these experiments; we have opted for a reduced training set in favour of including the smaller sections D, E, and F in our experiments, which were omitted from the PLuTO experiments. Again, for determining the relationship between the domains, we examine the best runner-up on each section, considering the BLEU score, although asymmetrical, as a kind of measure of similarity between domains. We can establish symmetric relationships between sections A and C, B and F as well as G and H, which means that the models are mutual runner-up on the other's test section. This shows that the same relationships can be inferred from BLEU scores as from the information theoretic measures evaluated before.

To ensure that these effects are not solely caused by the overlap between IPC classes, table 12 shows the relative pairwise document overlap between IPC sections, i.e. the percentage of documents in A that are also classified as B, C, D etc. in column A. We observe a large overlap of roughly 30% in both directions between A and C, accounting for the mutual runner-up result in the cross-section evaluation; the same holds for G and H. However, the amount of overlap is not the sole factor for the mutual translation performance on sections. C and B share over 20% of their documents, but C performs worst on B and vice versa. The relationship between sections is also not necessarily symmetrical: the smaller sections D, E, and F each share about 30% of their documents with B, which is mirrored in the translation score of B on all three sections where it is always runner-up. In the other direction, the influence of these sections on B is only small, but there seems to be a strong similarity between B and F which is

Table 11. BLEU scores for 300k individual IPC section models

| | test | | | | | | | |
train	A	B	C	D	E	F	G	H
A	*0.5349*	0.4475	**0.5472**	0.4746	0.4438	0.4523	0.4318	0.4109
B	0.4846	*0.4736*	0.5161	**0.4847**	**0.4578**	**0.4734**	0.4396	0.4248
C	**0.5047**	0.4257	*0.5719*	0.462	0.4134	0.4249	0.409	0.3845
D	0.47	0.4387	0.5106	*0.5167*	0.4344	0.4435	0.407	0.3917
E	0.4486	0.4458	0.4681	0.4531	*0.4771*	0.4591	0.4073	0.4028
F	0.4595	**0.4588**	0.4761	0.4655	0.4517	*0.4909*	0.422	0.4188
G	0.4935	0.4489	0.5239	0.4629	0.4414	0.4565	*0.4748*	**0.4532**
H	0.4628	0.4484	0.4914	0.4621	0.4421	0.4616	**0.4588**	*0.4714*

Table 12. Percentage document IPC overlap: $\frac{|X \cap Y|}{|Y|}$

| | Y | | | | | | | |
X	A	B	C	D	E	F	G	H
A	*100.0%*	9.7%	**31.5%**	15.4%	6.0%	5.9%	10.0%	2.5%
B	13.0%	*100.0%*	24.5%	**32.0%**	**23.6%**	**30.3%**	17.3%	11.7%
C	**36.2%**	**21.0%**	*100.0%*	28.4%	7.8%	8.2%	14.4%	8.9%
D	2.4%	3.8%	3.9%	*100.0%*	1.7%	1.5%	0.7%	0.5%
E	1.2%	3.5%	1.3%	2.1%	*100.0%*	5.0%	1.4%	0.9%
F	3.3%	12.6%	4.0%	5.2%	14.2%	*100.0%*	4.6%	5.0%
G	10.7%	13.8%	13.4%	5.0%	7.9%	8.9%	*100.0%*	**30.3%**
H	2.3%	8.0%	7.1%	2.9%	4.1%	8.3%	**26.0%**	*100.0%*

significantly stronger than the relationship of B to the other small sections. We conclude that the document overlap between two sections is an indicator but not the determining parameter for similarity and mutual translation performance.

Evaluation results for separately trained individual models across text domains are shown in table 13 and exhibit patterns similar to the evaluation of the IPC models. Again, each section is best translated with a model trained on data from the same section. The results on abstracts suggest that this section most strongly resembles the claims; the model trained on claims achieves a respectable score. On claims, abstract and description models yield an almost equal score, but the score drops substantially from the best result, supporting the notion that claims possess a very distinct structure and wording that is only captured by a model that is able to learn these characteristics from the data. With this data available, however, claims seem to be easiest to translate, yielding the highest overall BLEU score of 0.4879. On the other hand, all models score considerably lower on title data, which is no surprise considering the fact that titles consist only of noun phrases and translation quality depends highly on vocabulary coverage. Given the high type-token-ratio of this section, we discuss a method to expand coverage in the next section. The parallel data obtained from the descriptions presumably lacks in quality compared to the other sections due to its origin. Overall, the scores on descriptions are lower than on abstracts and claims but still higher than on titles. The abstract model again scores best on this section. Altogether, the abstract model seems to be the most robust and varied model, yielding the runner-up score on all other sections. The title model, in contrast, performs worst across all other sections. We attribute these results to the limited variety of grammatical structure observed in the training data – titles only consist of noun phrases – as well as the smaller amount of training data with regard to the absolute number of tokens.

Table 13. BLEU scores for 500k text domain models

train	test			
	title	abstract	description	claims
title	*0.3196*	0.2839	0.1743	0.3512
abstract	**0.2681**	*0.3737*	**0.2812**	**0.4076**
description	0.2342	0.32189	*0.3347*	**0.403**
claims	0.2623	**0.3416**	0.2420	*0.4879*

8   Compound Splitting on Textual Domains

We hypothesize that the high type-token-ratio on the German source side stems from a large number of compound words, due to the fact that inventions have to be described accurately and no proper names may be used. We therefore investigated the effect of German compound splitting as a preprocessing step on the text section dimensions. We trained and applied a compound splitting model on the German part of the training sets for abstract, claims and titles and the respective test sets. Predictably, the splitting raises the token count and average type frequency while lowering the type count.

We apply a simple empirical compound splitting method by [19], which is distributed as a script in the Moses toolkit. It considers all possible splittings of a compound word into known words, taking possible fillers, such as the s in Rotationsverdämpfer, and dropped letters into account. The vocabulary of known words is derived from a monolingual training corpus. The decision, if and how a word will be split, is based on word frequency estimates made on the same corpus. Given the word count in the corpus, the model picks the split S with the highest geometric mean of word frequencies of its parts p_i: $\text{argmax}_S (\prod_{p_i \in S} \text{count}(p_i))^{\frac{1}{n}}$.

This approach has the effect that if a compound word appears more frequently than its parts, it is left intact, ensuring that common compounds are not unnecessarily broken up. A phrase-based translation model will learn a correct translation even when an incorrect split is done consistently, if the compounds in the training data are split as well.

The effects of compound splitting for German-English translation are displayed in table 14. Compound splitting improves the score on all four sections but most substantially on the titles with an improvement of 0.0439 BLEU. This shows that the translation of titles is strongly influcenced by a large percentage of compound words and suffers mostly from a sparse data problem. Table 15 shows the influence of compound splitting on two translation samples. The splitting does not always result in a perfect translation, e.g. producing air jet instead of air nozzle, but it considerably reduces the out-of-vocabulary (OOV) rate.

Table 14. Compound splitting on training and test set

	BLEU score		OOV rate	
	raw	split	raw	split
title	0.3196	**0.3635**	10.00%	5.44%
abstract	0.3737	**0.3827**	3.35%	1.20%
description	0.3347	**0.3385**	3.90%	1.60%
claims	0.4879	**0.5022**	4.58%	2.42%

Table 15. Sample compound splitting on titles

source	Luftdüsenspinnmaschinen	Druckluftversorgungssystems
split source	luft düsen spinnmaschinen	druck luft versorgung systems
baseline system	OOV	OOV
compound split system	air jet spinning machine	compressed air supply system
reference	air nozzle spinning machine	compressed air supply system

9 Discussion

Statistical machine translation is highly dependant on the availability of sentence-parallel data for diverse domains. We documented a twofold approach to extract large amounts of parallel data from the MAREC patent corpus, namely a straight-forward method, where we align translated document sections, i.e. title, abstract and claims, and an indirect approach, where we find approximate translations of a whole document via the patent family id connection. The full statistics for the resulting German-English parallel corpus can be found in table 16, containing the number of unique parallel sentences and number of tokens on English and German side. The large amount of clean, parallel data consitutes a valuable resource for patent translation. Based on this, we plan to explore advanced topics such as cross-lingual patent retrieval.

We further analysed the corpus by exploring two different subdomain dimensions in terms of their relatedness, both with corpus similarits measures

Table 16. Full sentence-parallel corpus: number of unique sentences, number of tokens

	sentences	tokens en	tokens de
title	2,204,384	18,159,477	14,798,306
abstract	715,735	30,830,602	26,606,213
description	11,912,840	446,229,748	322,282,966
claims	8,181,791	496,421,795	431,568,084
total	23,014,750	991,641,622	795,255,569

and machine translation evaluation. We find that both the IPC and the document structure domains are well-delimitated and showed that pairwise domain similarity and translation performance correlate. Furthermore, we identify and adress one particular structural problem for German-English patent translation, namely the high type-token-ratio, which stems from the large amount of new terms produced by German compounding.

We plan a similar extraction of parallel data for French-English and French-German patent translation in the future. These resources will enable us to further identify and investigate topics in automated patent translation.

Acknowledgements. The authors would like to thank the Information Retrieval Facility (IRF) for providing the MAREC patent corpus.

This work was supported in part by DFG grant "Cross-language Learning-to-Rank for Patent Retrieval".

References

1. Wäschle, K., Riezler, S.: Structural and topical dimensions in multi-task patent translation. In: Proceedings of the 13th Conference of the European Chapter of the Association for Computational Linguistics, Avignon, France (2012)
2. Utiyama, M., Isahara, H.: A japanese-english patent parallel corpus. In: Proceedings of MT Summit XI, Copenhagen, Denmark (2007)
3. Gale, W.A., Church, K.W.: A program for aligning sentences in bilingual corpora. Computational Linguistics 19(1), 75–102 (1993)
4. Lu, B., Tsou, B.K., Zhu, J., Jiang, T., Kwong, O.Y.: The construction of a chinese-english patent parallel corpus. In: Proceedings of the MT Summit XII, Ottawa, Canada (2009)
5. Tinsley, J., Way, A., Sheridan, P.: PLuTO: MT for online patent translation. In: Proceedings of the 9th Conference of the Association for Machine Translation in the Americas (AMTA 2010), Denver, CO (2010)
6. Jochim, C., Lioma, C., Schütze, H., Koch, S., Ertl, T.: Preliminary study into query translation for patent retrieval. In: Proceedings of the 3rd International Workshop on Patent Information Retrieval (PaIR 2010), Toronto, Canada (2010)

7. Ceauşu, A., Tinsley, J., Zhang, J., Way, A.: Experiments on domain adaptation for patent machine translation in the PLuTO project. In: Proceedings of the 15th Conference of the European Assocation for Machine Translation (EAMT 2011), Leuven, Belgium (2011)
8. Braune, F., Fraser, A.: Improved unsupervised sentence alignment for symmetrical and asymmetrical parallel corpora. In: Proceedings of the 23rd International Conference on Computational Linguistics (COLING 2010), Beijing, China (2010)
9. Och, F.J., Ney, H.: A systematic comparison of various statistical alignment models. Computational Linguistics 29(1), 19–51 (2003)
10. Koehn, P.: Europarl: A parallel corpus for statistical machine translation. In: Proceedings of Machine Translation Summit X, Phuket, Thailand (2005)
11. Moore, R.C.: Fast and Accurate Sentence Alignment of Bilingual Corpora. In: Richardson, S.D. (ed.) AMTA 2002. LNCS (LNAI), vol. 2499, pp. 135–144. Springer, Heidelberg (2002)
12. Siegel, S., Castellan, J.: Nonparametric Statistics for the Behavioral Sciences, 2nd edn. MacGraw-Hill, Boston (1988)
13. Kilgarriff, A., Rose, T.: Measures for corpus similarity and homogeneity. In: Proceedings of the 3rd conference on Empirical Methods in Natural Language Processing (EMNLP-3), Granada, Spain (1998)
14. Ben-David, S., Blitzer, J., Crammer, K., Pereira, F.: Analysis of representations for domain adaptation. In: Proceedings of the 20th Annual Conference on Neural Information Processing Systems (NIPS 2006), Vancouver, Canada (2006)
15. Koehn, P., Hoang, H., Birch, A., Callison-Birch, C., Federico, M., Bertoldi, N., Cowan, B., Shen, W., Moran, C., Zens, R., Dyer, C., Bojar, O., Constantin, A., Herbst, E.: Moses: Open source toolkit for statistical machine translation. In: Proceedings of the ACL 2007 Demo and Poster Sessions, Prague, Czech Republic (2007)
16. Federico, M., Bertoldi, N., Cettolo, M.: IRSTLM: an open source toolkit for handling large scale language models. In: Proceedings of Interspeech, Brisbane, Australia (2008)
17. Heafield, K.: KenLN: faster and smaller language model queries. In: Proceedings of the EMNLP 2011 Sixth Workshop on Statistical Machine Translation (WMT 2011), Edinburgh, UK (2011)
18. Papineni, K., Roukos, S., Ward, T., Zhu, W.J.: Bleu: a method for automatic evaluation of machine translation. Technical Report IBM Research Division Technical Report, RC22176 (W0190-022), Yorktown Heights, N.Y. (2001)
19. Koehn, P., Knight, K.: Empirical methods for compound splitting. In: Proceedings of the 10th Conference on European chapter of the Association for Computational Linguistics (EACL 2003), Budapest, Hungary (2003)

Query Enhancement for Patent Prior-Art-Search Based on Keyterm Dependency Relations and Semantic Tags

Khanh-Ly Nguyen and Sung-Hyon Myaeng[*]

Department of Information & Communication Engineering
[*]Division of Web Science & Technology
Korea Advanced Institute of Science and Technology
Daejeon, 305-701, Republic of Korea
{lynk,myaeng}@kaist.ac.kr

Abstract. Prior art search is one of the most common forms of patent search, whose goal is to find patent documents that constitute prior art for a given patent being examined. Current patent search systems are mostly keyword-based, and due to the unique characteristics of patents and their usage, such as embedded structure and the length of patent documents, there are rooms for further improvements. In this paper, we propose a new query formulation method by using keyword dependency relations and semantic tags, which have not been used for prior art search. The key idea of this paper is to make use of patent structure, linguistic clues and use word relations to identify important terms. Moreover, to formulate better queries we attempt to identify what technology area a patent belongs to and what problems/solutions it addresses. Based on our experiments where IPC codes are used for relevance judgments, we show that keyword dependency relation approach achieved 13~18% improvement in MAP over the traditional *tf-idf* based term weighting method when a single field is used for query formulation. Furthermore, we obtain 42~46% improvement in MAP when additional terms are used through pattern-based semantic tagging.

Keywords: Patent Search, Patentability Search, Semantic Tags, Prior Art Search, Keyword Dependency Relation, Query Formulation.

1   Introduction

Patent analysis aims at obtaining relevant patents and analyzing them to produce patent maps [10] or discover trends ([3], [11], [12]). Patent search is often conducted by inventors, patent attorneys, patent examiners, and technical and business experts to find the prior art and mitigate the risks associated with their tasks. There are several ways in doing patent search such as: prior art or novelty search, validity search, infringement search, clearance search, etc. Prior art search is one of the most common search tasks with its goal being to find patent documents that constitute prior art to a given patent [5]. Prior art search is performed before filing an application to ascertain patentability of an invention, to determine novelty of the invention, and/or to invalidate a patent's claim of originality. The increasing amount and the growing need to access

M. Salampasis, B. Larsen (Eds.): IRFC 2012, LNCS 7356, pp. 28–42, 2012.

patent information require new methodologies and techniques to shorten search time and increase the quality of the search output to eventually increase the quality of the patents granted.

Successful automatic prior art search must pay attention to the quality of search queries generated from a patent document, because it may contain many ambiguous and vague terms. It is particularly difficult to formulate a successful query because patent writers tend not to use standard terms to make them look novel; relevant patents to be retrieved may not contain the terms used in the query patent. Retrieved patents containing the query terms, on the other hand, may not be relevant to the intent of the searcher. In addition, it is not easy to set the optimal query size [7]. Although a short query makes query processing fast, it can easily misrepresent the intent of the searcher. A query with many terms on the other hand is likely to contain noisy terms, not to mention that it requires excessive computing resource.

Noting the importance of formulating reasonable queries, this paper proposes a query formulation method for patent prior art search, which makes use of term dependency relations and semantic tags. We propose to use an algorithm to select the most representative terms based on keyword dependency relations. Furthermore, we use this algorithm in combination with semantic tags we attach automatically based on pre-constructed linguistic patterns. The tags reveal the field of invention, the problems and solutions being mentioned in a patent.

We report our results on experiments we ran with the NTCIR corpus. Instead of making traditional relevance judgments for individual query-document pairs, which is prohibitively expensive, we make use of the International Patent Classification (IPC) codes that have been assigned to individual patents in the test collection. As a result, the prior art search task is to retrieve patents whose IPC codes match those of the query patent at three different levels of abstraction.

The paper is organized as follows. Chapter 2 describes the background and some related works. Chapter 3 gives details of our method for query formulation, use of keyterm dependency relations for term weighting and semantic tagging-based automatic phrase extraction. Chapter 4 reports our experimental results and includes discussions. Lastly, we conclude with the major contribution of the paper and further work.

2   Background and Related Work

Since patents play an important role in Intellectual Property protection, there has been a growing interest in research into patent retrieval. NTCIR-3 [1] released patent test collections to enable researchers to systematically evaluate their methodologies. NTCIR-4 [2] started a new task for prior-art search, also called invalidity search, whose goal is to identify previously published patents in the collection for a given patent. Subsequent NTCIR workshops have continued and started patent-related tasks of different kinds.

Previous research on prior art search focused on methods of formulating queries by identifying keywords from patent documents based on some weighting schemes and selecting additional keywords for a high probability of retrieving relevant results. Keywords were extracted usually from Claims, which is considered the most informative field of a patent, to form a search query. To enhance the initial query, query expansion techniques were performed by extracting additional terms usually from the Description field.

Other variations exist. In [4], query terms were extracted from components in a topic claim and expanded by terms from explanation sentences related to the components in Detailed Description. [26] utilizes the rhetorical structure of a claim by applying an associative document retrieval method, in which a document is used as a query to search for other similar documents. To produce an initial query, the Claims field is segmented into multiple components and then used to search for candidate documents on a component by component basis. [27] uses two retrieval stages with query terms extracted from the Claims field. The query constructed from the Claims field is used to retrieve the top 1,000 patents at the first stage, which are then re-ranked by several techniques at the second stage. Other fields such as Title, and Abstract, as well as Claims an Description, were used to generate queries ([5] [8] [9]). While work in [18] shows that words from Titles are least useful for prior-art search, words from Titles are given preference in other work [17].

To evaluate the prior art search task, different collections have been constructed. In NTCIR-4, human expert judgments were used to generate relevance data for patent evaluation, but only 34 query topics were developed because of the cost. Also, in NTCIR-4 the IPC codes were integrated with a probabilistic retrieval model for estimating the document prior. In NTCIR-5 and NTCIR-6, citations were used and thousands of query topics were developed automatically. However, evaluations based on citations have some limitations because they have different degrees of relevancy and because citation lists are often incomplete [19]. Therefore, the IPC codes have been used as a feature for document filtering and patent retrieval. In [14] the authors use IPC codes for document filtering and show how this feature can help in patent retrieval.

3   Proposed Query Formulation Method

3.1   System Description

Our patent retrieval system consists of modules for Query Formulation based on keyword dependency relation and semantic tagging, Search, and Re-ranking. Query terms are extracted from Abstract, Claims and Description fields, and all redundant and stop-words are removed. Queries constructed with semantic tagging and term relation calculation are sent to Lemur to retrieve relevant patents.

Given the input patent, we segment it into sentences with punctuation marks. Unimportant terms are deleted from the input text fields using the van Rijsbergen's stop-word list [23] of 570 words and our own stop-word list of 150 words. We collected words, those occur frequently in patents but are not useful for the search (e.g. *figure, relates, said, apparatus, method, device*, etc.), manually from the patent documents during the patent analysis process. The total number of stop-words we used in this research is 720 words.

3.2   Term Annotation by Semantic Tags

The core content of a patent can be captured by a few key elements such as "What problem does the invention solve?", "What is the key invention?", and "What does the invention do?" [20]. The phrase representing the problem that an invention is going to solve is called *Problem* (P), and those representing what the invention does to solve the problems is called *Solution* (S). For example, "long-cycle-life lithium secondary cells" is the problem and "utilizing a lithium ionic reaction" is the solution. Problems and Solutions are important elements for describing the gist of a patent and hence a query. They help avoiding the need to construct lengthy queries. Note that Problems and Solutions can be shared within multiple patents in the same domain.

In addition, the *Invention Field* (IF) of a patent can help describing the area of technology (domain) which a patent belongs to (*e.g. secondary battery*). As in [17], patents would belong to the same domain if they share the same semantic tags which are defined by patent writers. In order to identify effective and concrete terms, we examine whether semantic tags such as *Invention-Field, Problems*, and *Solutions* (IFPS) are related to IPC codes which can aid in identification of related prior art patents. Together with *Problems* and *Solutions* tags, *Invention-Fields* can reduce the size of an input patent query so that searching can be done efficiently.

Figure 1 shows an example of IFPS phrases extracted from patents in the *batteries* domain and how IFPS phrases can assist in identifying the IPC codes that a patent belongs to. As in the figure, IF phrases such as "rechargeable *batteries*", "alkaline storage *batteries*", and "high power nickel metal hydride *batteries*" contain the word "*batteries*" which is the same as the name of IPC Sub-class (*Batteries*). Also the problem phrases such as "positive *electrode*", "positive *electrode* material" and "composite positive *electrode* material" contain the word "*electrode*" which is the same as the name of IPC Main-group (*Electrodes*). Similarly, solution phrases such as "*nickel* based multi metals oxide", "*nickel* hydroxide material", "composite *nickel* electrode hydroxide particulate" all contain "nickel" which is the same as the name of IPC Sub-group. Note that IF and problem and solution phrases do not always correspond to Sub-class, Main-group, and Sub-group, respectively.

The task of IFPS phrase extraction is to identify Invention Fields and Problem/Solution phrases from patent documents with the following two steps:

Step 1. Extraction of Phrases for Invention Fields (IF Phrase Extraction)

Though all patent documents have a similar structure and the titles of fields are fixed, the names of sub-fields are normally labeled by writers with no standard format (writer-defined tags). Since patents have separate *Invention Fields* with inconsistent titles such as "Field of the Invention" or "Technical Field" (Figure 2), automatically identifying the Invention Fields part of a patent document is not trivial. In some patents,

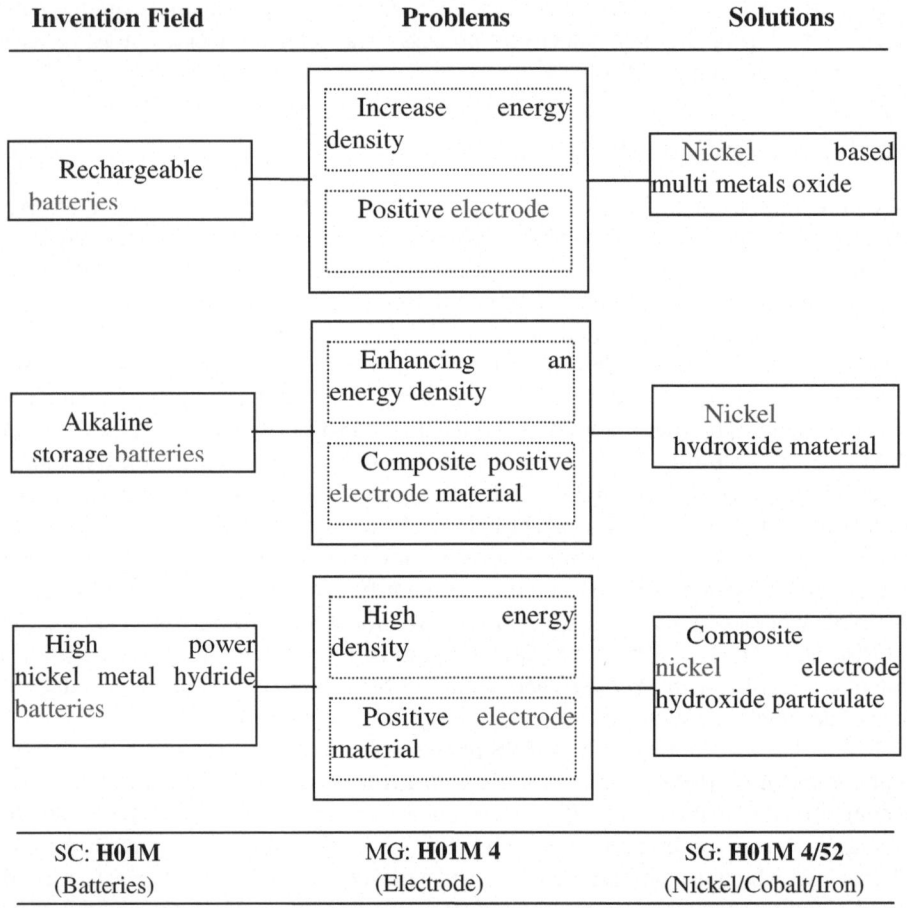

Invention Field	Problems	Solutions

Fig. 1. An example of relations between semantic tags and the IPC of H01M

furthermore, the text corresponding to the Invention Field is not separately marked but embedded in such fields as "Background of the Invention", "Prior art", "Description of the Related Art", etc (Figure 3).

For the cases with a separate writer-created field name for "Invention Field", which we collected for all the variations, we extract sentences under the field title. For the rest without such a field title (about 10% of the patents), we look for sentences that contain "relates to", which is a most likely clue for a description of "Invention Fields", under the field title like "Background of the Invention". Figure 3 shows an example where the first two sentences contain the clue phrase "relates to" under "Background of the Invention" that is under "Description". Note that we do not extract sentences using the clue from other fields such as Embodiments and Detailed Description that are not likely to describe the invention fields.

BACKGROUND OF THE INVENTION

1. Field of the invention

The present invention relates to a active material for positive electrode for an alkaline secondary cell and a method for producing the same as well as an alkaline secondary cell using the above active material for positive electrode, especially a nickel-hydrogen secondary cell and a method for producing the same.

2. Prior art

...

Fig. 2. An example of Invention Field under a writer defined tag

---

BACKGROUND OF THE INVENTION

The present invention relates to an active material used in a positive electrode for alkaline storage batteries such as nickel-cadmium storage battery, nickel-metal hydride storage battery and the like. More particularly, it relates to an active material comprising a nickel based multi-metals oxide.

Recently, an alkaline storage battery, particularly portable sealed storage...

Fig. 3. An example of Invention Field without an explicit writer-defined tag

Step 2. Extraction of Problem and Solution Phrases

Problems and Solutions are also extracted from the *Description* field. We first used Open NLP POS Tagger [22] to tag the text under Description fields and manually analyzed patent documents to generate a list of clues generally used repeatedly in a large number of patents. We utilized these linguistic clues to create 24 patterns[1] so that Problem and Solution phrases can be easily extracted through a simple pattern matching process. Figure 4 shows a sample of the patterns used for detection of Problem and Solution phrases.

After extracting IF and PS phrases, we formulate IFPS queries by combining IF and PS and removing all redundant and stop-words. Figure 5 shows an example of sentences that contain Invention Field, Problem and Solution phrases in italics.

3.3   Term Weighting by Keyterm Dependency Relations (KDR)

In addition to the semantic tagging for key phrase extraction for patent documents, we attempt to rank-order terms based on their mutual dependency in a field of technology.

---

[1] It is conceivable to generate the clues using a machine learning technique, which is left for future work. Instead, we constructed the patterns manually for high precision, which is essential for proof of concept.

The main idea is to increase the weight of a term that co-occures with other important terms that are likely to represent the domain of interest. In other words, we conjecture that a term used alone without a relationship with other content

Problem Sample Pattern
{NP}   +   can/MD   be/VB provided/VBN|improved/VBN|obtained/VBN
{NP} + is/VBZ improved/VBN in/IN + {NP}
apparatus/NN|methods/NNS for/IN + {NP|V-ing + NP}
provided/VBN + {NP} /

Solution Sample Pattern
utilizing/VBG|employing/VBG|using|VBG + {NP}
to/TO +{VBG+NP}

Fig. 4. A Sample of Problem and Solution Detection Patterns

Patent No.	Invention Field	Problems	Solutions
6177213	The present invention relates to an *alkaline storage battery*	a method for producing a *composite positive electrode material*	having *composite nickel hydroxide particulate*
6261720	The present invention relates to an active material used in a positive electrode for *alkaline storage batteries*	*high capacity positive electrode active material* for an alkaline storage battery	comprising a *nickel based multi-metals oxide*
6013390	The present invention relates to an improvement of *alkaline storage batteries*	Utilization of a *positive electrode active material* comprising a nickel oxide	comprising a *nickel oxide*
	SC: **H01M** (Batteries)	MG: **H01M 4** (Electrode)	SG: **H01M 4/52** (Nikel/Cobalt/Iron)

Fig. 5. An example for extracted IFPS

terms are likely to be an outlier and hence less helpful in identifying IPC codes for a quer patent.

Let us consider the following sentences, for example:

- "Thus, a **nickel**-metal hydride storage **battery** of high capacity can be provided."

- "*Nickel* based alloy layer for perpendicular recording media."

Since "nickel" and "battery" co-occur in the first sentence, "nickel" is likely to be in the "battery" domain[2]. In the second sentence, "nickel" co-occurs with other words but not with "battery". This is an indication that "nickel" is not related to the Battery domain.

We borrowed the idea from the method for building Keyword Dependency Profile [15] which utilizes keyword dependency relations (KDR) for topic tracking. The intuition is that a word may have strong dependency relations with other words, which is important for describing the main information. KDR is evaluated by co-occurrences among keywords in the same sentences. The initial weight of a keyword, which can be calculated by the *tf-idf* value, is increased if it is strongly related to other important keywords with a high initial weight.

Figure 6 is an example of a KDR graph where words come from the sentence "Thus, a **hydride** storage **battery** of high capacity can be provided". After removing stop-words and punctuations, we have a list of keywords <"hydride", "storage", "battery", "capacity", "provided">. The graph on the left shows the relationships and initial term weights calculated with *tf-idf*. The number on an edge is the frequency of the two words occurring together in the same sentences. After applying the KDR algorithm to be explained below, weights of the terms change as in the graph on the right. The more edges connected to important nodes, the higher the weight increases. For example, the weight of "hydride" is increased significantly since it connects to important nodes such as "capacity", "storage", and "battery".

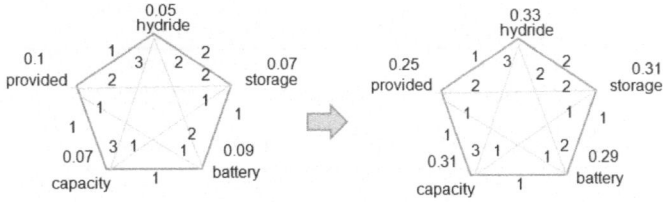

Fig. 6. An example of a KDR graph and weight changes

The KDR algorithm is applied as follows. An input text is first segmented into sentences. After removing all redundant and stop-words, each sentence is represented as a graph, in which each word is a node n, and each edge e is the connection between two nodes. The weight of each node $w(n_k)$ is re-calculated by the following formula:

$$w(n_k) = \sum_{l=1, l \neq k}^{m} \log(tf(n_l) \times idf(n_l) \times tf(e_{k,l}) + 1)$$

---

[2] We assume here that IF indicates that the patent is in Battery domain.

where m is the number of nodes that co-occur with node n_k in the same sentence, and l is the neighbor node that co-occurs with node n_k in the same sentence. Here $e_{k,l}$ is the edge that connects node n_k and node n_l, and $tf(e_{k,l})$ is the frequency of edge $e_{k,l}$ in the input text. A new weight is computed not only with $tf(n_l)$, term frequency of node n_l, and $idf(n_l)$, inverse document frequency of node n_l, but also with the frequency of the sentences containing the two terms in the entire input text.

3.4   Query Formulation

Given a patent, a base query is constructed by taking the top N terms based on the re-calculated weights from a field or combination of fields. There are three types of queries that we constructed in the retrieval process and evaluation: queries from an individual field, queries from combined fields, and queries with terms tagged by IFPS. Table 1 shows the details of the query sets we constructed and experimented with. We chose 10 and 60 terms from different fields empirically depending on their lengths.

3.5   Patent Indexing and Retrieval

The Lemur Indri search engine was used to index patent documents. No stemming or stop-word removal was done for the baseline implementation. For each query, we retrieved top 1000 patents from the corpus (its description explained in the next section), and each retrieved patent was assigned a relevant score. We used the Okapi BM25 formula for the ranking model in this retrieval which has been used in many retrieval experiments.

Table 1. Details of Experimental Query Sets

	Query Set	**Query Description**
Separate Field	Abstract	Top 10 terms from Abstract
	Claims	Top 10 terms from Claims
	Description	Top 60 terms from Description
Combined Fields	10 Abstract + 10 Claim	Top 10 terms from Abstract + Top 10 terms from Claims
	10 Abstract +60 Description	Top 10 terms from Abstract + Top 60 terms from Description
	10Abstract + 10Claim + 60Description	Top 10 terms from Abstract + Top 10 terms from Claims + Top 60 terms from Description
Merged with IFPS phrases and KDR	IFPS	IFPS phrases
	IFPS + 10Abstract	IFPS + Top 10 terms from Abstract (by KDR)
	IFPS + 10 Claim	IFPS + Top 10 terms from Claims (by KDR)
	IFPS + 10Abstract + 10Claim	IFPS + Top 10 terms from Abstract (by KDR) + Top 10 terms from Claims (by KDR)

4   Experiments and Results

4.1   Test Collection and Evaluation Measures

For experiments, three data sets were collected as follows: (1) A corpus of patent documents to search, (2) a set of patent queries, and (3) relevance judgments for patent documents in the corpus. For (1), we used the NTCIR-6 corpus consisting of 1,315,470 patent documents published from 1993 to 2002. All the fields of the patents (e.g. Title, Abstract, Claims, Description) were indexed using Lemur toolkit [13]. For (2), we crawled 244 patent documents belonging to the domain of Batteries (H01M) published from 2003 up to now from the USPTO patent search website [16]. For relevance judgments, which usually require a very large amount of human resources, we chose to use IPC codes. If the IPC codes of the retrieved patents match those of the query patent, we consider they are relevant to the query. An advantage of using IPC codes for semantic level matching between a query and a document is that IPC codes are at three different abstraction levels, Subclass, Main-Group and Sub-Group, in increasing order of specificity. It allows us to apply different weights to the three different levels of matching so that their relative influence on the retrieved ranked documents can be considered. Table 2 shows the statistics of relevant IPC codes for the patent query set. Tables 3 and 4 show the statistics of the queries formulated by applying KDR and semantic tags (IFPS), respectively. Note that about 10% of the patents do not have Invention-Field and only 1% of the patents have Problem/Solution phrases over 260 terms.

Table 2. Statistics of the relevant IPC codes

Relevant IPCs	Total	Min	Max
Sub-class	415	1	7
Main-group	1001	2	8
Sub-group	1829	6	23

Table 3. Statistics of the data extracted by the KDR method

Statistics of KDR Analysis	Average	Min	Max
#of graphs per query patent	40	6	256
#of nodes per graph	8	3	58
#of edges per node	7	2	57

Table 4. Statistics of Semantic tags: Invention Fields (IF) and Problems/Solutions (PS)

Statistics of IFPS	Average	Min	Max
#of IF per document	16	0	113
#of PS per document	53	7	263
#of IFPS per document	58	11	279

To measure retrieval effectiveness, we chose to use the most commonly used metrics in IR: Mean Average Precision (MAP) and Recall (R). The measures were computed with *trec_eval* program [21] which has been commonly used in the TREC evaluations.

4.2   Experimental Results

The main goal of the experiments is to examine the effects of the two proposed query formulation methods: using keyword dependency relations (KDR) and semantic tags including Invention Fields, Problems, and Solutions (IFPS). For comparison purposes, we chose the conventional *tf-idf* method in Lemur Indri search engine as our baseline since this is the most commonly used method in previous research in patent prior art search.

Query Formulation by Keyword Dependency Relation (KDR)

Our first interest was to examine our intuition that term weighting based on KDR would play an important role in selecting domain-sensitive terms in patent prior art in comparison with the *idf*-based weighting. The KDR-driven query formulation method outperformed *tf-idf* in the majority of the cases as in Table 5 where comparisons were made for individual fields: Abstract, Claims and Description. For the Abstract field, KDR shows a significant improvement in term of MAP with 18.2%, 17.1%, and 13.4% for Sub-class, Main-group, and Sub-group, respectively. The improvement for the Claims field is smaller with 7% for Sub-class and 5.5% for Main-group. The performance for Sub-group is slightly worse than *tf-idf*. For the case of using the Description fields, the KDR-based method has even smaller MAP improvements. This result is somewhat expected because KDR queries tend to select more common or general terms including acronyms that are more sensitive to the domain whereas the *tf-idf* based queries tend to have rarer terms in the collection. This tendency is illustrated in the following example.

-    Terms selected by KDR: *electrode, positive, nickel, oxide, temperature, ag, ca, material, effect, cr*
-    Terms selected by *tf-idf*: *charging, nickel, overvoltage, storage, alkaline, batteries, absorbing, positive, effect, oxygen*

Having observed that the absolute performance values in recall and MAP are in the order of Description > Claims > Abstract, we ran experiments to understand what would be the best performance we could obtain by mixing the fields. The best result was obtained when all the three fields were used together, compared to other combinations of two field types. Queries were created by selecting top 10 terms from Abstract, top 10 from Claims, and top 60 from Description. The numbers were chosen to reflect the relative sizes of the texts in the fields. The result is shown in Table 6 where it is clear that the KDR-based term weighting helps improving effectiveness in terms of recall and MAP across all the IPC code types. This result sets the upper bound with and without KDR.

Table 5. Results of two query formulation methods for individual fields

Field	Method	Sub-Class		Main-Group		Sub-Group	
		Recall	MAP	Recall	MAP	Recall	MAP
Abstract	tf-idf	**0.985**	0.566	**0.928**	0.442	0.768	0.206
	KDR	0.983 (-0.2%)	**0.669** (**+18.2%**)	0.916 (-1.3%)	**0.517** (**+17.1%**)	**0.824** (**+7.3%**)	**0.234** (**+13.4%**)
Claims	tf-idf	**0.988**	0.640	**0.945**	0.499	**0.840**	**0.237**
	KDR	0.986 (-0.2%)	**0.685** (**+7.0%**)	0.942 (-0.3%)	**0.527** (**+5.5%**)	0.830 (-1.1%)	0.228 (-3.7%)
Descrip -tion	tf-idf	0.995	0.775	0.962	0.592	0.909	**0.289**
	KDR	**1.000** (**+0.5%**)	**0.808** (**+4.3%**)	**0.966** (**+0.4%**)	**0.609** (**+2.8%**)	**0.910** (**+0.1%**)	0.289 (-0.1%)

Query Formulation by Semantic Tags (IFPS)

We ran experiments to see the effect of semantic tagging with IFPS over the *tf-idf* method. The same number of terms was extracted for both cases from the Description fields, which gave the best results. As in Table 7, we gained some improvements across all the IPC code types. The biggest gain was observed for Sub-class, mainly because the terms extracted from IFPS are at a fairly general level.

Given that the IFPS-based query formulation rank terms that are related to the major areas of invention and the problems and major solutions, we attempted to see their effect on the queries extracted from more succinct text, i.e. Abstract and Claims. Table 8 shows that the improvement in MAP over the *tf-idf* method is close to 50% for all the IPC code types when query terms were extracted by both KDR and IFPS from the Abstract fields only. Note that the improvement in recall with Sub-group is quite significant. We suspect that when the terms extractable from the patent text are at a fairly general level, the IFPS terms as well as the KDR method are very effective in improving retrieval effectiveness.

As can be seen in Table 9, medium level improvements were obtained using a combination of KDR and IFPS when the Claims fields were used for the retrieval experiment. While the percent increases are less than those in Table 8, the trend is similar to the case of using the Abstract fields: marginal and significant improvements for recall and MAP, respectively.

Table 6. Queries formulated from a combination of Abstract, Claims and Description

Field	Method	Sub-Class		Main-Group		Sub-Group	
		Recall	MAP	Recall	MAP	Recall	MAP
Top10 from Abstract + Top 10 from Claims + Top 60 from Description	tf-idf	0.995	0.767	**0.964**	0.589	0.097	0.291
	KDR	**0.998** (+0.3%)	**0.818** (+6.7%)	**0.964** 0%	**0.622** (+5.5%)	**0.914** (+0.8%)	**0.298** (+2.3 %)

Table 7. IFPS compared with *tf-idf* when the same number of words was extracted from the Description field

Method	Sub-Class		Main-Group		Sub-Group	
	Recall	MAP	Recall	MAP	Recall	MAP
tf-idf	**0.998**	0.772	0.962	0.593	0.914	0.289
IFPS	**0.998** (+0%)	**0.831** (+7.6%)	**0.972** (+1%)	**0.619** (+4.3%)	**0.918** (+0.4%)	**0.294** (+1.7%)

Table 8. Effects of queries obtained from KDR and IFPS when succinct text is used for query formulation

Method	Sub-Class		Main-Group		Sub-Group	
	Recall	MAP	Recall	MAP	Recall	MAP
10 Abs (Tf-idf)	0.985	0.566	0.928	0.442	0.768	0.206
10 Abs (KDR) + IFPS	**0.998** (+1.3%)	**0.831** (+46.8%)	**0.970** (+4.5%)	**0.631** (+42.6%)	**0.921** (+19.9%)	**0.300** (+45.3%)

Table 9. Effects of queries obtained from KDR and IFPS with the Claims fields

Method	Sub-Class		Main-Group		Sub-Group	
	Recall	MAP	Recall	MAP	Recall	MAP
10 Claims (tf-idf)	0.988	0.640	0.945	0.499	0.840	0.237
10 Claims (KDR) + IFPS	**0.995** (+0.7%)	**0.843** (+31.7%)	**0.972** (+2.9%)	**0.634** (+26.9%)	**0.921** (+9.6%)	**0.302** (+27.5%)

4.3   Overall Summary

We carried out experiments for query formulation by two methods: keyword dependency relations (KDR) and semantic tags (IFPS). Queries were extracted by taking the top N number of terms from each field or a combination of two or three fields. Retrieval results were evaluated for three IPC codes, Sub-class, Main-group and Sub-group, in comparison with those obtained by the *tf-idf* method. The experimental results show that Description is the most useful field for query formulation compared with Abstract or Claims. However a combination of the top N terms from the fields gave better performance than any of the fields in separation. KDR and IFPS gave better performance than *tf-idf* while the best performance was obtained when KDR and IFPS were used together. IFPS-based query formulation was particularly useful when shorter or more succinct text in Abstract or Claims fields were used in formulating queries. Moreover, we found that for each IPC class type, terms from a different field would give different results in IPC class identification.

5   Conclusion

The main contribution of the paper is its proposal to use term dependency relations and semantic tagging for query formulation in prior art search in the patent retrieval domain. The KDR-based term weighting method and IFPS phrase extraction method were shown to be effective in improving MAP for the task of identifying the IPC codes at three different levels of abstraction. KDR tend to help identifying domain-specific terms whereas IFPS helps finding terms that are essential in judging the key invention and the problems being tackled in a patent. As such, they are most effective in identifying the Sub-class IPC codes, followed by Main-group and Sub-group.

We also found in our experiments, that the Description fields gave the best performance in single field experiments although most of the previous research used terms from Claims for query formulation. It is because Description field is more lengthy and contains more useful terms, especially terms that relates to the main gist of a patent such as Problems, Solutions and Invention Field, that help in the patent's IPC identification. It was observed that when the three fields, Abstract, Claims, and Description, are combined, the overall best performance was obtained, regardless of the term weighting or extraction schemes we employed.

Another valuable insight we obtained from this work is that the traditional term weighting method based on tf-idf has an inherent weakness. Since the frequency of the terms often found in Invention Field tends to be low, they cannot be extracted purely based on frequency information. Some of the problem and solution phrases have the same tendency.

While the proposed methods can be applied to patent documents in different domains, we plan to run a larger scale experiments to cover the domain other than "Battery". Furthermore, the manually constructed patterns for IFPS phrase identification can be automated with machine learning techniques although the sub-language aspect of patent documents tend to contain many predictable clues for identifying Invention Fields, Problems, and Solutions. An immediate extension to this work is to consider customizing the KDR algorithm for patent document such as using IFPS in computing term dependency relations.

References

[1] Iwayama, M., Fujii, A., Kando, N., Takano, A.: Overview of patent retrieval task at NTCIR-3. In: Proceedings of NTCIR Workshop (2009)

[2] Fujii, A., Iwayama, M., Kando, N.: Overview of Patent Retrieval Task at NTCIR- 4. In: Proceedings of NTCIR-4 Workshop (2004)

[3] Kim, Y., et al.: Automatic Discovery of Technology Trends from Patent. In: Proceedings of the 2009 ACM Symposium on Applied Computing, pp. 1480–1487 (2009)

[4] Kazuya, K.: Query Term Extraction from patent documents for invalidity search. In: Proceedings of NTCIR-5 Workshop Meeting, Tokyo, Japan, December 6-9 (2005)

[5] Roda, G., Tait, J., Piroi, F., Zenz, V.: CLEF-IP 2009: Retrieval experiments in the Intellectual Property domain. In: CLEF-IP (2009)

[6] Susan, V., Eva, D.: Prior Art retrieval using the claims section as a bag of words. In: CLEF-IP (2010)

[7] Toucedo, J.C., Losada, D.E.: University of Santiago de Compostela at CLEF-IP09. In: 1st CLEF-IP, Corfu, Greece (2009)

[8] Xiaobing, X., Bruce Croft, W.: Transforming Patents into Prior Art Queries. In: SIGIR 2009 (2010)

[9] Metti, Z., et al.: Prior art retrieval using various patent document fields contents. In: CLEF-IP (2010)

[10] Mai, F.-D., Hwang, F., Chien, K.-M., Wang, Y.-M., Chen, C.-Y.: Patent map and analysis of carbon nanotube. Science and Technology Information Center, National Science Council, ROC (2002)

[11] Young Gil, K., et al.: Visualization of patent analysis for emerging technology. Expert Systems with Applications: An International Journal archive 34(3) (April 2008)

[12] Lent, B., et al.: Discovering trends in text databases. In: Proc. 3rd Int. Conf. Knowledge Discovery and Data Mining, KDD, pp. 227–230 (1997)

[13] The Lemur Toolkit, http://www.lemurproject.org

[14] Takaki, et al.: Associative Document Retrieval by Query Subtopic Analysis and its Application to Invalidity Patent Search. In: Proceedings of CIKM (2004)

[15] Zheng, W., Zhang, Y., Hong, Y., Fan, J., Liu, T.: Topic Tracking Based on Keywords Dependency Profile. In: Li, H., Liu, T., Ma, W.-Y., Sakai, T., Wong, K.-F., Zhou, G. (eds.) AIRS 2008. LNCS, vol. 4993, pp. 129–140. Springer, Heidelberg (2008)

[16] The USPTO databased, http://www.uspto.gov/

[17] Kim, J.-H., et al.: Patent document categorization based on semantic structural information. Information Processing and Management (2007)

[18] Xue, X., Bruce Croft, W.: Automatic Query Generation for Patent Search. In: Proceeding of the 18th ACM Conference on Information and Knowledge Management, CIKM 2009 (2009)

[19] Lupu, M., Mayer, K., Tait, J., Trippe, A.J.: Current Challenges in Patent Information Retrieval. The Information Retrieval Series 29 (2011)

[20] Hunt, D., Nguyen, L., Rodgers, M.: Patent searching: tools & techniques (2007)

[21] trect_eval program at TRECT website, http://trec.nist.gov/trec_eval

[22] Open NLP POStagger, http://opennlp.sourceforge.net/

[23] van Rijsbergen, C.J.: Information Retrieval, 2nd edn. Butterworths, London (1979), http://www.dcs.gla.ac.uk/Keith/Preface.html

Analysis and Detection
of Web Spam by Means of Web Content

Víctor M. Prieto, Manuel Álvarez, Rafael López-García, and Fidel Cacheda

Department of Information and Communication Technologies,
University of A Coruña, Campus de Elviña s/n. 15071 A Coruña, Spain
{victor.prieto,manuel.alvarez,rafael.lopez,fidel.cacheda}@udc.es
http://www.tic.udc.es/

Abstract. Web Spam is one of the main difficulties that crawlers have
to overcome. According to Gyöngyi and Garcia-Molina it is defined as
"any deliberate human action that is meant to trigger an unjustifiably
favourable relevance or importance of some web pages considering the
pages' true value". There are several studies on characterising and detecting Web Spam pages. However, none of them deals with all the possible
kinds of Web Spam. This paper shows an analysis of different kinds of
Web Spam pages and identifies new elements that characterise it. Taking them into account, we propose a new Web Spam detection system
called SAAD, which is based on a set of heuristics and their use in a C4.5
classifier. Its results are also improved by means of Bagging and Boosting techniques. We have also tested our system in some well-known Web
Spam datasets and we have found it to be very effective.

Keywords: Web characterization, Web Spam, malware, data mining.

1 Introduction

Currently, the WWW constitutes the biggest repository ever built, and it is con-
tinuously growing. Due to its large size, search engines are essential for users who
want to access relevant information. According to a study by Jansen and Spink
[16], approximately 80% of search engine users do not take into consideration
those entries that are placed beyond the third result page. The increasing use of
search engines (Google, Yahoo!, Bing, etc.), has made companies and web de-
velopers worry about the ranking of their Web sites. In order to achieve the first
positions, it is important to improve the quality of the Web sites and to renew
their contents and their "relationships" with web sites about similar topics.

Nevertheless, achieving a good place in the ranking is neither trivial nor inex-
pensive, so some techniques known as Web Spam [11] appeared with the purpose
of getting an unfair relevance of the Web pages or sites. There are many people
and organisations that are interested in using Web Spam to damage third parties
(usually competitors) or to increase the PageRank in order to achieve a better
position and raise their income for publicity or link selling. In fact, there are Web
Spam organisations with qualified personnel and the purpose of earning money

M. Salampasis and B. Larsen (Eds.): IRFC 2012, LNCS 7356, pp. 43–57, 2012.
© Springer-Verlag Berlin Heidelberg 2012

illegally, too. Among their most common techniques (which can be combined to make their detection more difficult), we have:

- Content Spam, that is a technique based on the modification of the content or the keywords with the purpose of simulating more relevance to search engines and attract more traffic.
- Cloaking, which consists in dynamically generating different content for certain clients (e.g.: browsers) but not for others (c.g.: crawling systems).
- Redirection Spam, which consists in hiding redirections to pages with different content by means of scripts. Some clients (e.g.: browsers) will follow these redirections, but the rest (e.g.: crawlers) will not detect them, obtaining only the static content that the user will never see.
- Link Spam, which is the creation of Web Spam by means of the addition of evil links between pages with the purpose of raising their popularity. It is also possible to create "link farms", which are pages and sites interconnected among themselves with the same purpose.

Because of all the things we explained before, we should create protection mechanisms to a) the final users who can be cheated and who are wasting their time and money, b) the companies or the owners of the pages that want to get clients by means of ethical methods and c) the companies that provide search engines. The latter are very affected since they do not only lose prestige when they show Web Spam pages among their results, but they are wasting money and resources in analising, indexing and showing results of pages that should not be shown. Not protecting some of these entities means an economic loss.

Taking this into account, we propose SAAD (Spam Analyzer And Detector), a system that uses web content to identify Web Spam pages. A crawling system could incorporate this module in order to avoid Web Spam pages and, more concretely, in order not to analyse, index or show them to the final user.

The structure of this article is as follows. In section 2 we summarize the work on Web Spam techniques and their detection. Section 3 shows the presence of Web Spam on the Web and establishes the need of detecting it. It also discusses the properties we use in our system to characterise Web Spam. Section 4 explains our method for Web Spam detection, which is based on the combination of different heuristics. Section 5 shows the features of the datasets we have chosen to assess the detection methods, as well as the results we have obtained in every one of them. We also compare our results to previous experiments. Finally, in sections 6 and 7 we comment our conclusions and future work respectively.

2   Related Work

Although Web Spam has existed since the Web appeared and it has been growing with the expansion of the Web, its study in the academic sphere is recent.

Some important articles study Web Spam in general, like the one by Henzinger *et al.* [14], who discuss the importance of the phenomenon and the quality of the results that search engines offer. Gyöngyi and Garcia-Molina [11] propose a taxonomy of Web Spam, too. The survey by Castillo and Davison [4] shows further reference on this topic. However, most of them focus on some of the main Web Spam types: Content Spam, Cloaking, Redirection Spam and Link Spam.

Regarding Content Spam, Fetterly et al. [8], Ntoulas and Manasse [18] and Gonzalez et al. [10] highlight the importance of analysing the content and their properties in order to detect Web Spam. Fetterly et al. [9] also discuss the special interest of the "cut-and-paste" content between Web Pages in order to detect Web Spam. Hidalgo [15] and Sahami et al. [22] propose similar techniques for Email Spam detection, combining content analysis and classifiers (e.g.: C4.5).

With regard to Cloaking, Gyöngyi and Garcia-Molina [11] explain some existing techniques. Wu and Davison [25] propose a detection method which is based on calculating the common words between three copies of a web page.

Link Spam has been studied by Wu and Davison [26] and Gyöngyi and Garcia-Molina [12], who presented some methods for detecting link farms. Amitay et al. [1] also analyse the importance of the properties of connectivity among pages in order to identify them. On the other hand, studies like the one by Zhang et al. [29] propose methods to prevent link farms from affecting the PageRank. Benczur et al. [2] introduce the idea of TrustRank, which consists in starting from a set of "clean" pages that have been analysed previously, extracting links from them and adding the targeted pages with the appropriate level of trust.

Finally, Redirection Spam has been studied by Chellapilla and Maykov [5] and Wu and Davison [25], who presented analyses about the use of this technique.

There are other techniques related to Web Spam that aim to attack browsers by means of evil JavaScript (Malware Spam). These techniques employ illicit actions such as "drive-by-downloads" attacks to take control of the system. Once the machine is in control, the attacker can perform several kind of attacks, like Web Spam, Email Spam or even DDOS (Distributed Denial Of Service) attacks. Cova et al. [6] study the most common techniques and their detection.

The articles we have mentioned deal with the different Web Spam techniques and, in some cases, they present their corresponding detection methods. The problem of those studies is that they treat Web Spam detection partially, since they focus exclusively on some of the described types but not in their combination. We propose a new method to detect the different kinds of Web Spam based on existing techniques as well as in the new techniques we have developed.

3   Detection of Spam by Means of Content Analysis

The importance of Web Spam on the Web is obvious. Ntoulas et al. [18] have created a random dataset of several domains, deciding manually whether those pages were spam. According to this study, about 70% of the pages of the .biz domain are Web Spam, followed by 35% of the .us domain. Domains like .com, .de and .net have between 15% and 20% of Web Spam pages. Although this amount is lower, it still remains high. On the other hand, we have the .edu domain, whose pages seem to be completely free from Web Spam. The study remarks that a high percentage of the whole Web is Web Spam, so an appropriate detection that prevents systems from dealing with those pages is important.

In the next sections, we first summarize the set of existing Web Spam detection heuristics. We will discuss both the ones based in content analysis and the others. After this, we will analyse the set of detection heuristics we propose.

3.1   Detection of Some Kinds of Web Spam

Systems as the ones proposed by Ntoulas *et al.* [18] and Gonzalez *et al.* [10] base their detection on content analysis, applying the following heuristics:

- Number of words per page: they are counted and compared to the typical ratios since common words are often added to the page to raise its relevance.
- Number of words in the title: it is the same as the previous one, but applied to the title of the page since it is usually considered a relevant element.
- Word length: it measures the average length of the words, since it has been found that small words are often joint to make longer ones (e.g.: "freemp3").
- Anchor words: it calculates a ratio between the length of the text of the "anchor" tags and the length of the document, since spammers often use this text to hide links that point to spam pages.
- Ratio of the visible content: this method calculates the ratio between the content without HTML tags and the total content. This way we can know if spammers are trying to hide Web Spam in the non-visible texts.
- Compression ratio: the size of the page itself is compared to a compressed version to know whether spammers are trying to repeat certain words.
- Number of common words: spammers usually add common words from queries to capture more traffic. To avoid it, we calculate the ratio between the N most common words and the total number of common words.
- Independent n-gram probability: since gramatical and semantical analysis are very expensive, this heuristic proposes an statistical analysis of the content to detect Web Spam. To that end, we divide the content of each page in n-grams of n consecutive and independent words and compare their probabilities to the ones of Web Spam pages.
- Dependent n-gram probability: the same as the previous one but introducing dependencies between words.

3.2   SAAD Heuristics for Web Spam Detection

The techniques we have explained in the previous section only enable detection of Web Spam in pages which use Content Spam techniques, but our goal is to deal with all the types of spam we commented before (Cloaking, Redirection Spam, Content Spam and Link Spam). With this purpose we have performed an analysis in which we have identified new heuristics.

From here on, we discuss thoroughly the main heuristics on which SAAD is based. In order to justify the efficacy of each one, we tested them on the dataset we describe in section 5. In each figure, we show the spam probability according to the values that we have obtained for each technique. We show the ratio of pages that are affected by each value of each heuristic too. In order to improve the presentation of the data, we have used a log10 progression in the Y axis that indicates this ratio, but showing the percentage instead of the absolute value.

- Word length II: we propose an improvement of the method shown in the previous section, which consists in calculating the average length of the content without taking into account the HTML tags or the stop words. We do this because HTML tags are not content that is going to be shown to the user and would introduce noise in the results, and also because we have observed that no-spam pages employ a bigger amount of stop words. This happens because a legitimate page uses a normal quantity of prepositions, conjunctions or articles, but spam pages mainly focus on the insertion of keywords that improve their position in the search rankings. In Figure 1a we can observe that a big portion of the pages contain words with a length between 2.5 and 7.5. We can also say that the probability for a page of being spam is below 25% if its average length (without HTML tags or stop words) is lower than 8. However, if the values are bigger we can see that the probability increases progresively until it reaches near 90%. If we compare to the results obtained by Ntoulas et al. [18], we can see that we obtain more conclusive results, since from a value of 8, the progression grows and has better probabilities.

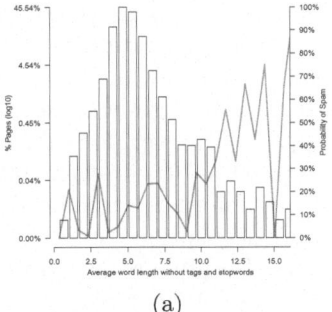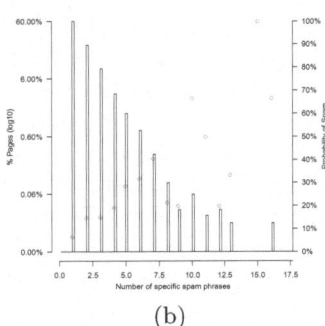

(a)   (b)

Fig. 1. Probability of spam relative to the average word length II of the pages (a), and to specific phrases in the page content (b)

- Specific phrases: it is usual that spam pages contain common terms, phrases or queries from search engines. This has also been observed in other areas like Email spam where emails usually contain common sentences or specific spam words. Analysing our dataset, we have created a list of the aforementioned terms, which contains words like "viagra" or "urgent" among others. In Figure 1b we observe the results. On the one hand, there are lots of pages with low number of spam words and a low probability of being Web spam. On the other hand, the number of pages with many spam words is lower and its probability of being Web Spam is, in general, higher.
- Encode/decode functions: spammers often try to hide redirections to other pages, functions or a certain content by codifying it. Hence, the use of functions like escape/unescape is usual. It has also been found a combination of

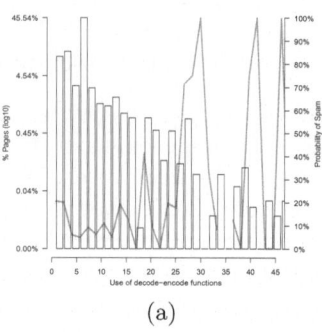
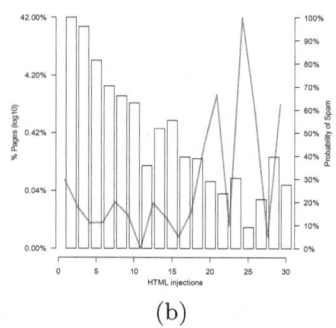

(a)   (b)

Fig. 2. Probability of spam relative to encode/decode functions in the page (a), and probability of spam relative to HTML injections in the page (b)

those functions one inside another to make their detection much more difficult. In Figure 2a it can be observed that the spam probability is relatively low if there are between 0 and 17 functions, so it is more probable that those pages are not spam. Nevertheless, between 17 and 22 functions and from 25 on the Web Spam probability grows reaching points with 100%.

- HTML injection: although most of the HTML is generated from user actions, we have observed that it is bigger in pages with spam. Hence, we have analysed the scripts in order to find functions that generate that code, such as: "innertext", "outerhtml", "createElement", "appendChild", etc. The results are shown in Figure 2b, where we observe that pages with less than 15 HTML injections can be considered as no-spam. However, for values bigger than 15 the probability of being spam rises to 60%, 70% or even 100%.

- Number of Keywords/Description Words: we performed an analysis of the number of words used in the "keywords" and "description" attributes of the META tag, in spam and no-spam Web pages. Results are shown in Figure 3a, where we can see that pages with less than 130 keywords have less of 20% of probability of being spam. When the number rises, the probability of being Web Spam grows as well. Taking this results into account, we decided to use more heuristics. Not only the "keywords" and "description" attributes of the META tag, but also the number of occurrences of these keywords in the attributes and through the page content.

- Images without Alt: since the content of many Web Spam pages is dynamically generated or is not thoroughly detailed, we have analysed the content of the ALT attribute of the images. Figure 3b points out that with values between 0 and 25, the probability of spam is less than 10%. If the number is bigger, we have peaks up to 60%. When the number is bigger than 220, the probability is bigger than 50%, with some peaks of 100%.

Apart from the heuristics we have studied, we have defined another set of heuristics. These are non-determining, but significant for the classification. We highlight the following:

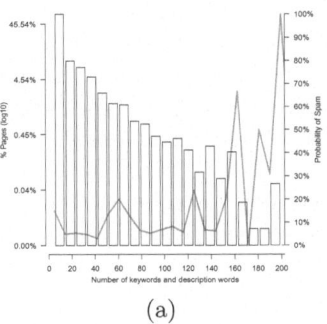
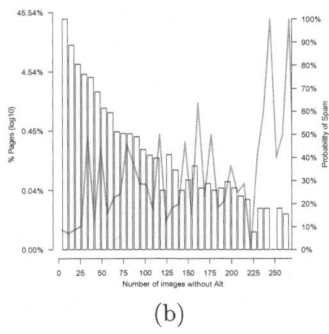

(a)   (b)

Fig. 3. Probability of spam relative to number of keywords and description words in the page (a) and to images without ALT tag in the page (b)

- Ratio of bytes of code and total bytes: during the analysis of the pages, we noticed that the ratio between the size in bytes of the scripting code of a page and the total number of bytes grows according to the probability of spam. The bigger the code in the HTML, the bigger the probability of spam.
- Length of the evaluated code: it has been observed that the evaluated strings are longer than usual in Web Spam pages, since in many cases they contain a large number of escaped variables, functions and data.
- META tag redirection: on the one hand, spammers put content in the current page in order to make crawlers analyse and index it, but, on the other hand, they redirect users to pages with different content. We have also detected that redirections in the META tag of Web Spam pages, often have a delay of less than 5 seconds.
- Orthography: Web Spam pages often generate content automatically or by means of "copy&paste". This is why, as well as in Email Spam, the rate of mistypes is bigger than in a no-spam page.
- Number of images: spam pages are often generated automatically, so it is unusual that they have images or at least they usually do not have a number of images similar to the one of no-spam pages. Taking this into account, we measure the total number of images of each page.
- Size in bytes of the Web page: we have checked that many spam pages have a size lower than the average of normal pages in the studied datasets.
- Number of stop words: a common indexing method for search engines is extracting the stop words from the Web page and then indexing it. Hence, spammers usually decide to include incoherent content, without conjunctions, articles, prepositions or other common words, this is, without stop words. As a technique to detect this fact, we propose to analyse the amount of stop words in the content (without HTML tags).
- Popular words: in section 3.1 we mentioned the detection of Web Spam by measuring the most common words of the content by means of different methods. We also propose to focus exclusively in the keywords and description, since they are the places where the use of common words is bigger, so spammers use them to get more traffic for their web site.

Finally, we have also taken into account heuristics that are based on theoretical features of the Web Spam pages, such as the presence of a high amount of hiddent text; the massive use of redirections, scripting functions, dynamic invocation of functions, activeX and so on. However, we have not found these heuristics as effective as we expected (and less than the ones we submitted previously), so we will not discuss them more thoroughly.

As we said before, some of the previous techniques can be used for licit purposes, although our idea is to prove that their intensive use and their combination is a determining indicator of Web Spam pages.

4   Method for Web Spam Detection

In this section we describe SAAD, the proposed method for Web Spam detection. As in Ntoulas et al. [18] and Gonzalez et al. [10], it is based on content analysis. However, we take into account the heuristics we have defined in section 3.2 in order to be able to detect all the known types of Web Spam: Cloaking, Link Spam, Redirection Spam and Malware Spam, as well as Content Spam.

For the appropriate combination of the heuristics, we have tried different classification techniques (decision trees, techniques based on rules, neural nets, etc.). We conclude that we achieve the best results when we use decision trees. More concretely, we have chosen the C4.5 algorithm [21] [20].

In order to improve the results, we have assessed two techniques, "bagging" [3] [19] and "boosting" [19]. These techniques create a set of N classifiers, combine those that obtain the best results and build a composite classifier. The "bagging" technique creates N subsets of n random elements with replacement. This way, N classifiers are obtained. Each Web page that wants to be classified has to be evaluated by each one of the N classifiers. The class in which the Web page will be added (spam or no-spam) depends on the votes of most of the N classifiers. The "boosting" technique works in a similar way. Each item has a weight associated. This weight reflects the probability of occurrence in the set. N classifiers are generated in N iterations, but for each misclassified item its weight is increased. Once again, the final decision of marking it as spam or not will be brought by a majority of results of the N classifiers.

Figure 4 shows a portion of the decision tree generated to classify a page as spam. Each node uses one of the heuristics presented in section 3.2 and, according to different thresholds, decides or delegates to another node to refine the classification.

5   Experimental Results

In this section we first explain the features of the datasets we used to obtain the results. Then, we explain different matters we have taken into account for the execution and result extraction. In the last two subsections, we thoroughly show and analyse the results we have obtained for each dataset.

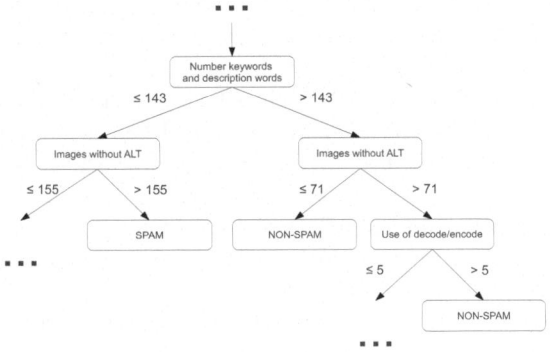

Fig. 4. A portion of decision tree

5.1 Dataset

In order to test the techniques we have used two well-known Web Spam datasets:

- Webb Spam Corpus: created by Webb *et al.* [23], [24], it is the biggest Web Spam dataset, having more than 350,000 different spam pages. The authors used Email Spam detection techniques to detect spam emails. Then, they followed the links that those emails proposed and they stored the Web pages.
- WEBSPAM-UK2006/7: this dataset was created by Yahoo! [28], whose collection has more than 100 million Web pages from 114,529 hosts, although only 6,479 were manually tagged. The spam percentage is 6%. It was created for the "Web Spam Challenge 2008" [27].

The first dataset only contains Web Spam pages. To add no-spam pages, we have randomly mixed this dataset with no-spam pages from the Yahoo! one. Due to the quality and quantity of the data that these datasets provide, the conclusions that we have obtained are very reliable. Another feature we would like to highlight, apart from the size, is that both datasets are public, in contradistinction to the datasets used in similar articles. This means that any researcher can check and assess the methods and results we show by means of a similar experiment.

5.2 Experimental Setup

In order to perform the experiments, we have developed a tool that allows us to execute and store the result for each heuristic we have explained before and for each Web page we have analysed from the mentioned datasets. This tool also stores the processing time of each one. This data have allowed us to check which heuristics perform better and which ones take less processing time. The tool is also in charge of filtering each page of the datasets, preventing us from analysing pages with no content, minimal content or with only JavaScript redirections.

In order to choose the most appropriate algorithm for our heuristics, we have employed WEKA [13], a tool for automatic learning and data mining, which includes different types of classifiers and different algorithms for each classifier. Once the heuristics had been chosen, our experiment has been executed and we have obtained different results for many classification algorithms. After analysing these results, we have chosen the C4.5 classification algorithm as the best one.

For the evaluation of the classifier we have used "cross validation" [17], which consists in building k data subsets. In each iteration a new model is built and assessed, using one of the sets as "test set" and the rest as 'training set". We have used "ten-fold cross validation" ($k = 10$), since it is a widely used number.

Our experiments have been divided in two types depending on the dataset we have used. First, we have applied the heuristics proposed by Ntoulas et al. [18] and Gonzalez et al. [10] in the "Webb Spam Corpus" dataset. Then, these results have been compared to the ones obtained by our system and to the ones obtained by extending our heuristics with the ones proposed by Ntoulas et al. In the case of the Yahoo! dataset, we have compared the results obtained by our system to the results of the winners of the Web Spam Challenge 2008.

5.3   Results in Web Spam Dataset of Webb

We have evaluated SAAD with several datasets with 22,760, 100,000, 150,000 and 200,000 documents respectively. After this, the results have been used first as an input for the C4.5 algorithm and then they have been applied Bagging and Boosting techniques. Figure 5a shows the results obtained for dataset 1 with 22,760 pages (2,760 spam pages and 20,000 no-spam pages). This size is similar to the one employed by Ntoulas et al. [18]. Figure 5b shows the ones obtained for the dataset 2 with 100,000 pages (50,000 spam and 50,000 no-spam). The results we have obtained in the rest of datasets with different sizes are consistent with the ones discussed before, so we do not show them.

For the results we have obtained using only the C4.5 algorithm, we can see that SAAD improves the recall obtained by Gonzalez et al. by approximately 10% in both datasets. In the case of Ntoulas et al., SAAD improves their results by 7% in dataset 1 (Figure 5a) and 12.5% in dataset 2 (Figure 5b). If we focus on the union of the heuristics submitted by Ntoulas et al. and the ones we have proposed (SAAD and Ntoulas row), we can see that we improve an additional 1.5%. In the case of the precision, we improve the results for dataset 1 by approximately 3% (Figure 5a).

After applying Bagging, SAAD improves about 10% the results by Gonzalez et al. for dataset 1 and 8.2% for dataset 2 (Figure 5b). SAAD also achieves an improvement of 5.6% in the results by Ntoulas et al. for dataset 1 and 10% for dataset 2. In this case, SAAD obtains 4% better results than the union of our heuristics and the ones proposed by Ntoulas et al.

The results after applying Boosting are better than the previous ones. Once again, SAAD improves the previous results. For dataset 1, SAAD achieves an improvement of 6% in the results submitted by Ntoulas et al. and Gonzalez et al. For dataset 2, we improve the results by 6% in the case of Gonzalez et al.

	Spam		No-Spam		Spam		No-Spam	
	Precision	Recall	Precision	Recall	Precision	Recall	Precision	Recall
C4.5								
Gonzalez et al.	0.895	0.715	0.962	0.988	0.977	0.66	0.742	0.984
Ntoulas et al.	0.894	0.743	0.965	0.988	0.967	0.641	0.705	0.975
SAAD	0.91	0.813	0.975	0.982	0.96	0.766	0.797	0.969
SAAD and Ntoulas	0.915	0.826	0.976	0.996	0.936	0.715	0.744	0.944
C4.5 - Bagging								
Gonzalez et al.	0.958	0.719	0.962	0.979	0.99	0.669	0.749	0.993
Ntoulas et al.	0.954	0.761	0.968	0.995	0.981	0.65	0.712	0.986
SAAD	0.969	0.815	0.975	0.996	0.982	0.75	0.797	0.986
SAAD and Ntoulas	0.968	0.828	0.977	0.99	0.962	0.712	0.747	0.969
C4.5 - Boosting								
Gonzalez et al.	0.932	0.79	0.972	0.992	0.985	0.703	0.768	0.989
Ntoulas et al.	0.942	0.791	0.993	0.949	0.974	0.657	0.719	0.98
SAAD	0.972	0.85	0.978	0.997	0.983	0.767	0.818	0.987
SAAD and Ntoulas	0.982	0.851	0.979	0.998	0.949	0.776	0.81	0.958
	(a)				(b)			

Fig. 5. Results for dataset 1 (22,760 pages, 2,760 spam and 20,000 no-spam) (a), and results for dataset 2 (100,000 pages, 50,000 spam and 50,000 no-spam) (b)

and approximately by 10% in the case of Ntoulas *et al.* If we focus on the results obtained by joining our heuristics with the ones proposed by Ntoulas *et al.*, we find that we only improve by 1% the results obtained if we only use our system.

In Figure 6 we show the best results we have obtained. With the information from this figure and from Figure 5a and Figure 5b, we can observe two different tendencies. First, the results are better if we use C4.5 and Bagging, and they improve even more if we use Boosting. Second, the best results are obtained when we join the techniques we have proposed before and the ones submitted by Ntoulas *et al.* [23], although in some cases the results are the same or worse than the ones obtained by applying only SAAD. Finally, we have the ones obtained by applying the methods proposed by Ntoulas *et al.* and Gonzalez *et al.* respectively.

5.4   Results in Web Spam Dataset of Yahoo!

In the previous section there were only two classes of pages: spam and no-spam. In the case of Yahoo! dataset, a new "undecided" class has been created. This class contains those pages which are in the limit between spam and no-spam. Before commenting the results, we want to discuss some matters that will help to understand them. First, we will only show the result of the C4.5 algorithm with Boosting, since in the previous section it has been stated that it is the technique that produces the best results. Second, we have not analysed the results obtained

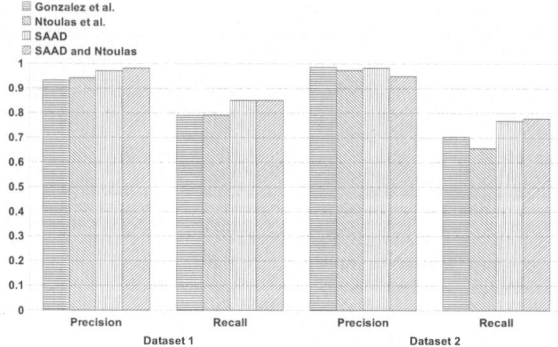

Fig. 6. Summary of results

by Ntoulas *et al.* and Gonzalez *et al.* in the Web Spam Challenge [27], since it has been proved in the previous section that our methods obtain better results. Finally, we show and compare the ROC area since it is the method used by Yahoo! for showing the results.

We have created two subsets from the Yahoo! dataset, the first one with 20,000 Web pages and the second one with 50,000. Both datasets have been created by means of a random selection and always maintaining the 6% of spam that Yahoo! indicated.

Dataset	20000				50000			
	Class	Precision	Recall	ROC	Class	Precision	Recall	ROC
	Spam	0.992	0.966	0.997	Spam	0.999	0.971	0.996
C4.5 - Boosting	No-Spam	0.99	0.999	0.991	No-Spam	0.996	0.999	0.996
	Undecided	0.985	0.87	0.989	Undecided	0.986	0.953	0.995

Fig. 7. Results on Yahoo! datasets

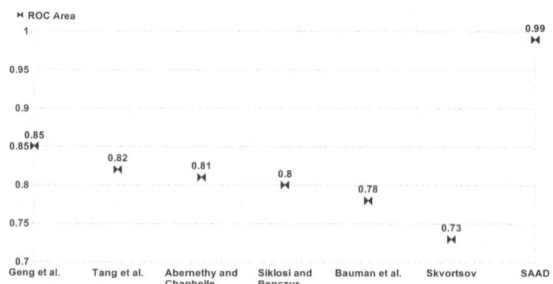

Fig. 8. Comparison of results by ROC area

In Figure 7 we can observe the results obtained for both datasets. The first, with 20,000 pages, has 17,679 no-spam pages, 1,270 spam pages and 1,043 Undecided pages. The second dataset we show has 50,000 pages, containing 43,584 no-spam, 3,368 spam and 3048 Undecided. The results we have obtained are very good, since in the spam detection SAAD obtains a ROC area between 0.996 and 0.997. Similar results, near 0.99, have been obtained for the no-spam and Undecided classes. Figure 8 shows clearly the improvements for the results we have obtained with respect to the results obtained in Yahoo!'s Web Spam Challenge. In spam detection, SAAD has achieved an improvement of 15% in the worst case and near 27% in the best case. The results for no-spam detection or Undecided pages have not been shown by Yahoo!, so we cannot make a comparison.

6   Conclusions

We have proposed a wide number of properties that help us to detect Web Spam and that improve and increase the ones that have been discussed in other articles. Our heuristics do not deal with the relationships between Web pages or sites, but we analyse a lot of features of the Web content to try to detect Web Spam.

Each technique we have presented could identify by itself a Web Spam page, but legitimate ones could be also classified as spam. In the same way, if we trusted in the techniques independently, spammers could avoid them easily. For this reasons, we have proposed the union of all the heuristics in order to create a more robust detector. This improves a lot the probability of success when we are detecting Web Spam and makes it more difficult to avoid all techniques.

Our methods have been tested with two public datasets. First, we have used one created from the Email Spam (Webb Spam Corpus), which exclusively contains Web Spam pages. Then, we hvae used a Yahoo! dataset, manually tagged.

At the same time we have joined the methods, we have created a tool that generates the results. These results have been used as the input of several classifiers. We have proved that the best results are obtained by means of a C4.5 classification algorithm improved by applying Boosting. With this, we improve from 6% to 10% the results submitted by Ntoulas *et al.* and Gonzalez *et al.*

In the case of the Yahoo! dataset [28], we have achieved a ROC area up to 0.99, this is, between 15% and 27% better than the results achieved by the winners of the Web Spam Challenge 2008 [27].

To the best of our knowledge, these results are the best we have found in similar studies, so the use of the proposed system in crawlers would generate a safer client environment, saving resources since they would not waste them in analysing, indexing or even showing Web Spam.

7   Future Work

We contemplate a future analysis of the profits and efficiency of including a module in web browsers, in order to alert the user when risky pages are being shown. We also think about using it in a crawling system, so it will improve the resource use by not wasting them in analysing and indexing pages that are potential Web Spam. With this, we would improve the ranking results shown to the user. We would also like to discuss the possibility of generalising the Web Spam detector to a distributed architecture, based in the Map-Reduce technique[7], which is currently the most used in crawling environments.

Another complementary study would be the analysis of the computing times of each method and the possibility of combining them in two or more stages, separating those which have lower resource use and computing time, and those which are more effective in the detection but they are also very resource demanding. This way, the system would be more efficient, since if a first stage with a low resource use marks a page as Web Spam, it would not be necessary to execute other resource demanding modules.

Acknowledgments. This work was partialy supported by the Spanish government, under project TIN 2009-14203.

References

1. Amitay, E., Carmel, D., Darlow, A., Lempel, R., Soffer, A.: The connectivity sonar: detecting site functionality by structural patterns. In: Proceedings of the Fourteenth ACM Conference on Hypertext and Hypermedia, pp. 38–47. ACM Press (2003)
2. Benczur, A.A., Csalogany, K., Sarlos, T., Uher, M., Uher, M.: Spamrank - fully automatic link spam detection. In: Proceedings of the First International Workshop on Adversarial Information Retrieval on the Web (AIRWeb) (2005)
3. Breiman, L., Breiman, L.: Bagging predictors. In: Machine Learning, pp. 123–140 (1996)
4. Castillo, C., Davison, B.D.: Adversarial Web Search 4(5), 377–486 (2010)
5. Chellapilla, K., Maykov, A.: A taxonomy of javascript redirection spam. In: Proceedings of the 3rd International Workshop on Adversarial Information Retrieval on the Web, AIRWeb 2007, pp. 81–88. ACM, New York (2007)
6. Cova, M., Kruegel, C., Vigna, G.: Detection and analysis of drive-by-download attacks and malicious javascript code. In: Proceedings of the 19th International Conference on World Wide Web, WWW 2010, pp. 281–290. ACM, New York (2010)
7. Dean, J., Ghemawat, S.: Mapreduce: simplified data processing on large clusters. Commun. ACM 51, 107–113 (2008)
8. Fetterly, D., Manasse, M., Najork, M.: Spam, damn spam, and statistics: using statistical analysis to locate spam web pages. In: Proceedings of the 7th International Workshop on the Web and Databases: colocated with ACM SIGMOD/PODS, WebDB 2004, pp. 1–6. ACM, New York (2004)
9. Fetterly, D., Manasse, M., Najork, M.: Detecting phrase-level duplication on the world wide web. In: Proceedings of the 28th Annual International ACM SIGIR Conference on Research & Development in Information Retrieval, pp. 170–177. ACM Press (2005)
10. Gonzalez Jesus, B.W., Cristina, A.: Implementacion y evaluacion de un detector masivo de web spam (2009)
11. Gyongyi, Z., Garcia-Molina, H.: Web spam taxonomy. Technical Report 2004-25, Stanford InfoLab (March 2004)

12. Gyöngyi, Z., Garcia-Molina, H.: Link spam alliances. In: Proceedings of the 31st International Conference on Very Large Data Bases, VLDB 2005, pp. 517–528. VLDB Endowment (2005)
13. Hall, M., Frank, E., Holmes, G., Pfahringer, B., Reutemann, P., Witten, I.H.: The weka data mining software: an update. SIGKDD Explor. Newsl. 11, 10–18 (2009)
14. Henzinger, M.R., Motwani, R., Silverstein, C.: Challenges in web search engines. SIGIR Forum 36, 11–22 (2002)
15. Hidalgo, J.M.G.: Evaluating cost-sensitive unsolicited bulk email categorization (2002)
16. Jansen, B.J., Spink, A.: An analysis of web documents retrieved and viewed (2003)
17. Kohavi, R.: A study of cross-validation and bootstrap for accuracy estimation and model selection, pp. 1137–1143. Morgan Kaufmann (1995)
18. Ntoulas, A., Manasse, M.: Detecting spam web pages through content analysis. In: Proceedings of the World Wide Web Conference, pp. 83–92. ACM Press (2006)
19. Quinlan, J.R.: Bagging, boosting, and c4.5. In: Proceedings of the Thirteenth National Conference on Artificial Intelligence, pp. 725–730. AAAI Press (1996)
20. Quinlan, J.R.: Improved use of continuous attributes in c4.5. Journal of Artificial Intelligence Research 4, 77–90 (1996)
21. Quinlan, J.R.: C4.5: programs for machine learning. Morgan Kaufmann Publishers Inc., San Francisco (1993)
22. Sahami, M., Dumais, S., Heckerman, D., Horvitz, E.: A bayesian approach to filtering junk e-mail (1998)
23. Webb, S.: Introducing the webb spam corpus: Using email spam to identify web spam automatically. In: Proceedings of the 3rd Conference on Email and AntiSpam (CEAS) (2006) (Mountain View)
24. Webb, S.: Webb Spam Corpus (2011), http://www.cc.gatech.edu/projects/doi/WebbSpamCorpus.html
25. Wu, B., Davison, B.D.: Cloaking and redirection: A preliminary study (2005)
26. Wu, B., Davison, B.D.: Identifying link farm spam pages. In: Special Interest Tracks and Posters of the 14th International Conference on World Wide Web, WWW 2005, pp. 820–829. ACM, New York (2005)
27. Yahoo!: Web spam challenge (2011), http://webspam.lip6.fr/wiki/pmwiki.php
28. Yahoo!: Web Spam Detection - Resources for Research on Web Spam (2011), http://barcelona.research.yahoo.net/webspam/
29. Zhang, H., Goel, A., Govindan, R., Mason, K., Van Roy, B.: Making Eigenvector-Based Reputation Systems Robust to Collusion. In: Leonardi, S. (ed.) WAW 2004. LNCS, vol. 3243, pp. 92–104. Springer, Heidelberg (2004)

Discovery of Environmental Nodes in the Web

Anastasia Moumtzidou[1], Stefanos Vrochidis[1], Sara Tonelli[2], Ioannis Kompatsiaris[1],
and Emanuele Pianta[2]

[1] Informatics and Telematics Institute, Thessaloniki, Greece
{moumtzid,stefanos,ikom}@iti.gr
[2] FBK, Trento, Italy
{satonelli,pianta}@fbk.eu

Abstract. Analysis and processing of environmental information is considered of utmost importance for humanity. This article addresses the problem of discovery of web resources that provide environmental measurements. Towards the solution of this domain-specific search problem, we combine state-of-the-art search techniques together with advanced textual processing and supervised machine learning. Specifically, we generate domain-specific queries using empirical information and machine learning driven query expansion in order to enhance the initial queries with domain-specific terms. Multiple variations of these queries are submitted to a general-purpose web search engine in order to achieve a high recall performance and we employ a post processing module based on supervised machine learning to improve the precision of the final results. In this work, we focus on the discovery of weather forecast websites and we evaluate our technique by discovering weather nodes for south Finland.

Keywords: Environmental, domain specific search, keyword spices, classification, key-phrase extraction.

1 Introduction

Environmental data and measurements are considered of utmost importance for humanity, as they are strongly related to health issues (e.g. allergies), as well as to a variety of important activities (e.g. agriculture). In everyday life, environmental information is also of particular interest for outdoor activities such as sports and trip planning. Such activities are directly affected by weather conditions, air quality and pollen concentration, which comprise the most important environmental aspects. Nowadays, such environmental measurements are performed by stations established by environmental organisations and the measured data are usually made available through web portals and services. Unfortunately, the existing services for retrieving environmental information are not many and in addition most of them are not publicly accessible. Therefore, the main sources of such information are indisputably the web portals, sites and pages. The latter constitute the nodes that provide environmental information and therefore are mentioned as "environmental nodes". In order to support people in everyday action planning considering the environmental conditions,

M. Salampasis and B. Larsen (Eds.): IRFC 2012, LNCS 7356, pp. 58–72, 2012.
© Springer-Verlag Berlin Heidelberg 2012

we need to provide them with services that combine and orchestrate complementary and competing environmental information from several nodes, with a view to generating more reliable environmental measurements. More specifically, based on the literature on environmental forecasts, it is in general acceptable that ensembling the measurements of various providers and forecast models yields more precise results. The first step towards this direction is the automatic discovery of environmental nodes, which is addressed by this paper.

The discovery of environmental nodes can be considered as a domain-specific search problem. However, in this particular case, the problem is more complicated, since the discovery is oriented towards particular environmental web pages, (i.e. sites that contain environmental measurements e.g. weather forecasts, air quality observations) and not towards a more broad set of environmental websites not containing measurements (e.g. environmental scientific articles), which a normal domain-specific search engine could retrieve. To this end we combine state-of-the-art techniques of domain-specific search, advanced textual processing of websites and supervised machine learning with a view towards discovering environmental nodes. Specifically, we apply the two main domain-specific techniques, the keyword spice (i.e. domain-specific terms) generation [1], [2] and domain-specific query submission to a general-purpose search engine to achieve a high recall performance and therefore retrieve a satisfactory number of environmental nodes to cover the need of having information from multiple providers. Then we employ a post processing module based on supervised machine learning to improve the precision of the results. Achieving high precision is also of great importance, since content extraction techniques will be applied on the relevant sites in order to extract interesting information (i.e. measurements). In this work, we focus on the discovery of weather forecast websites and we evaluate our technique by discovering weather nodes for south Finland.

The contribution of this paper is the empirical study on the environmental sites, the development of an environmental-specific search engine, as well as the methodology for domain-specific search, which combines state-of-the-art techniques of domain-specific search and novel website supervised classification, based on key-concept extraction tools. To the best of the authors' knowledge, this is one of the first attempts to develop an environmental-specific search engine.

This paper is structured as follows: section 2 presents the relevant work, while section 3 describes briefly the type of environmental web pages we attempt to retrieve. Section 4 introduces the proposed architecture of the environmental node discovery, while sections 5 and 6 present the main components involved. The experiments and results are presented in section 7 and finally, section 8 concludes the paper.

2   Related Work

Although this work deals both with web site retrieval and domain-specific search we will present the prior work of the latter area, since this is the main focus of the paper.

In general, domain-specific search refers to the discovery of information relevant to a given domain by "domain-specific search engines". These search engines attempt to index only web pages that correspond to a pre-defined area, or topic. The main methodologies used for the implementation of a domain-specific search engine contain several techniques such as web searching, web crawling and query expansion. The proposed methodologies can be divided into two categories. The first is based on existing web search engines to retrieve a first set of results, which are subsequently minimized with the aid of post processing techniques. The second category is based on a predefined set of web sites and attempt to expand it using directed crawling.

In the approaches of the first category existing general-purpose search engines (e.g. Google Search, Yahoo! Search) can be adopted and the two following query generation approaches are used. In the first approach, a domain oriented query is generated by applying machine learning techniques in order to extract terms (called keyword spices) from positive and negative sample web pages. This extended query is then forwarded to a general-purpose search engine [1], [2]. The technique of keyword spices is described in more detail in section 5.2. An alternative methodology to generate such terms [3] employs the notion of context (i.e. domain) in order to automatically construct queries. The query creation mechanism is based upon entity-typed patterns and the expansion is performed using co-occurring terms that are produced from the analysis of previous user queries. In the second approach, the results obtained from the general-purpose search engines by submitting queries consisting of empirical terms that describe the domain, are filtered with filtering models [4] and post-analysis of the retrieved information [5], [6]. Specifically, Shakes et al. [4] forward user queries to a general-purpose search engine and shift out the irrelevant documents through the use of domain-specific filters that are based on empirical rules for identifying typical patterns found in the interesting pages. In [5], the authors have demonstrated that web content mining clustering techniques involving Self-Organizing Map algorithms can be applied to perform post-retrieval analysis of a retrieved document set, which generally improves the searching experience. Finally, in a recent approach [6], the authors attempt to automatically enrich an ontology by post-processing domain-specific search. The first step involves the automatic generation of queries using the concepts of an ontology. Then, these queries are submitted to a general-purpose search engine and the results are filtered through an SVM classifier. Our work integrates and combines both approaches in a common framework with a view of exploiting their advantages and eliminating their drawbacks.

The second category of methods is based on domain focused crawlers. Focused crawlers are able of retrieving web pages in a directed fashion using machine learning techniques. Several methodologies have been proposed for the implementation of a focused crawler. In [7], [8] the authors used reinforcement learning to build a domain-specific search engine. In another work, a focused crawler was implemented based on a machine learning approach [9], in which a supervised classifier was applied. In [10] a focused crawler was proposed, which was based on hypertext classification that tries to classify documents by exploiting only the link and not the content information. Finally, in [11] a learnable focused crawling framework based on ontology was proposed based on Artificial Neural Network. The main disadvantage of focused crawling approaches compared to the search engine based ones, is that it is difficult to achieve high recall, since this will require crawling billions of websites, while the works of the first category exploit huge existing web indices.

3   Empirical Study of Environmental Websites

In this section, we describe briefly the type of environmental web pages that we aim to discover (focusing on weather websites), their special traits and how the information is encoded. In order to identify the characteristics of environmental related web pages, we conducted an empirical study based on the empirical cycle methodology, which included the following phases: observation, induction, deduction, testing and evaluation. After performing an empirical study on more than 60 environmental nodes (from which about 20 focusing explicitly on weather), we conclude that the key characteristics of these sites are that they provide measurements for weather phenomena such as temperature, humidity and wind instead of general information (e.g. articles discussing environmental issues) and in addition, they include geographical location (e.g. city names) and time information (e.g. days). The interesting information found in such sites is usually encoded in textual format and therefore the proposed approach is text-based. An example of such websites is depicted in Figures 1 and 2. While in several sites (especially the ones providing air quality and pollen measurements), useful information is also presented in the form of images, maps and graphics, in this paper we focus on text-based environmental information, since it is considered adequate for addressing the websites that present weather measurements.

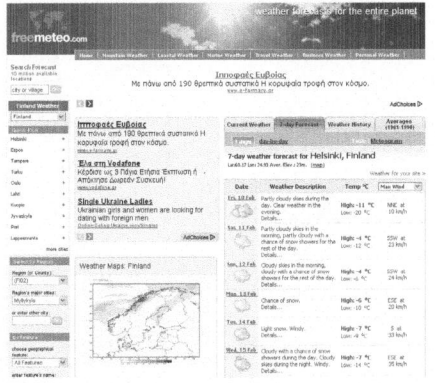

Fig. 1. Screenshot of typical example of weather site (freemeteo)

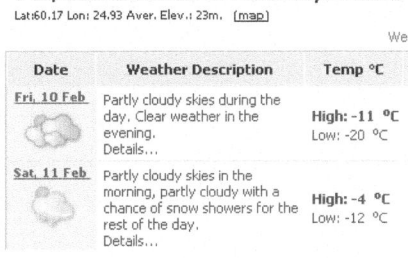

Fig. 2. Part of the "freemeteo" site containing interesting information

4   General Framework

The architecture proposed draws upon the first category of methodologies that have been described in section 2, which are based on general-purpose search engines. The use of such techniques is preferred because they are based on existing well-known and effective search engines that incorporate large indices built from a significant part of the web. Otherwise, the employment of crawlers would have required continuous indexing and local storage of large indexing structures.

The idea is to employ a hybrid architecture that combines the results from both methodologies of this category in order to exploit their advantages. More specifically, we provide the general-purpose search engine with both basic queries, as well as extended queries using keyword spices. The results of both techniques are then filtered through a supervised filtering post processing procedure based on advanced content analysis techniques, in order to remove part of the irrelevant sites and therefore improve the precision of the system. Post-processing analysis includes advanced extraction of textual features from websites taking into account the language and the structure of the webpage. The architecture of the system is depicted in Figure 3.

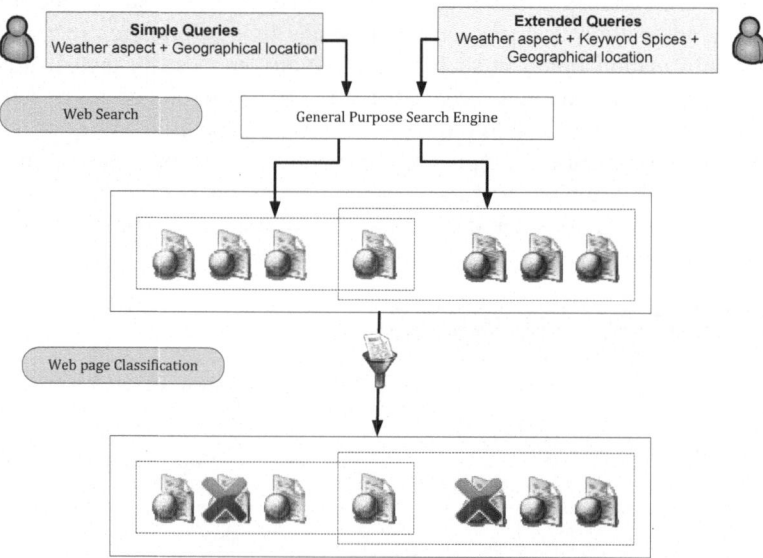

Fig. 3. Weather site discovery architecture

5   Web Search

The first step (i.e. web search) of the architecture presented in section 4 comprises the formulation of two types of domain-specific queries: the basic and the extended.

5.1   Basic Queries

The formulation of basic queries is based upon the conclusions of the empirical study (section 3). Therefore, the query (qb_i) is produced by combining the members w_i of the weather aspect set W with the members g_i of the geographical location set G, which includes city names. Therefore the basic query format is the following:

$$qb_i = w_i + g_i$$

An example of such a query is: "weather+Helsinki". Using this pattern, a set of queries can be formulated resulting from the combination of weather aspects with geographical regions that are of interest and are sent to the general-purpose search engine. Based on the empirical study, as weather aspects we consider the following set of weather related phenomena measured in the majority of the weather sites:

$$W = \{temperature, sky\ condition, pressure, rain, humidity, wind\}$$

It should be noted that the selection of weather aspects could be also done in an automatic way using existing lexical resources or ontologies. The city selection for any region could be also performed automatically by employing free geographical databases such as GeoNames[1].

5.2   Extended Queries

In addition to the aforementioned basic queries, we generate extended queries by generating domain-specific keywords that would be characteristic of the websites we look for. As discussed in section 2, keywords spices are domain-specific terms submitted to general-purpose search engines [1], [2]. This is achieved by extending the initial query with a domain-specific Boolean expression of keywords, called "keyword spices" (*ks*). This technique comprises two main steps: the identification of *ks* and their simplification. The first step includes the manual selection of a sample of pages, their classification into two classes T (relevant to the domain) or F (irrelevant to the domain) and their text-processing in order to extract their nouns that will be used as keywords. In the sequel, the retrieved web pages are split in a random way into two disjoint subsets, one used for identifying the *ks* and the other for simplifying them. Finally, a decision tree is created using the first set of web pages that uses the information gain measure. The output of the decision tree is an extensive Boolean expression from keywords/ nouns. Figure 4 depicts part of the decision tree output.

(forecast AND instrument AND animation)
OR (forecast AND ^instrument AND factor AND tracker AND screen)
OR (forecast AND ^instrument AND factor AND tracker AND ^screen)
OR (forecast AND ^instrument AND ^ factor AND radar AND ^government AND community AND ^cause)

Fig. 4. Part of the decision tree output

Given that general-purpose search engines can accept limited number of keywords as input, the second step involves the simplification of the original Boolean expression. During this step, an iterative procedure is performed including: i) the removal of

---

[1] http://www.geonames.org/

keywords that result in the increase of the harmonic mean of each conjunction and ii) the removal of conjunctions that result in the increase in the harmonic mean of the original disjunctive expression. The harmonic mean HM depends on precision P and recall R and is computed using the following formula: $HM = \{\frac{2}{\frac{1}{P}+\frac{1}{R}}\}$. The final result is a simple Boolean expression that contains keywords that should be included or not in the retrieved pages. Therefore, the extended query (qe_i) is produced by combining the basic query qb_i with the keyword spices (ks_D), that is:

$$qe_i = qb_i + ks_D$$

6   Site Post Processing

Subsequently, we describe the post processing procedure applied to the weather sites retrieved in the first step of retrieval using the general-purpose search engine. The advantage of the "web search" techniques described in the previous section is the ability to retrieve many relevant results (i.e. achieving high recall) by exploiting the web index of the general-purpose search engines. However it is highly possible that many irrelevant web pages will be retrieved as well, especially by the basic queries. For this reason, we introduce a post processing step that functions as a filter, which removes the irrelevant sites and improves the precision performance. This is achieved with the aid of a Support Vector Machine (SVM) classifier, which is trained with textual features to identify the relevant sites.

6.1   Classification

In the classification process, a SVM classifier is employed, due to the fact that SVMs have been applied successfully on several relevant classification problems. SVMs perform classification by constructing hyperplanes in a multidimensional space that separate cases of different class labels [12]. The procedure followed for creating an SVM model, that incorporates these hyperplanes, is a two step procedure that includes the building of the model and its testing.

1. *Model building/training*: The training of the model comprises of two main steps:
   (a) the definition of the features to be used and generation of a feature vector and
   (b) the selection of the parameters for the model.
2. *Model testing*: During testing, the created models are tested with new data and results are evaluated.

In this implementation, we employed the LIBSVM [13] library and we consider a binary C-Support Vector Classification using as kernel the radial basis function:

$$K(f_i, f_j) = e^{-g|f_i - f_j|^2}$$

where f_i and f_j are feature vectors of input data i, and j and g is a constant parameter.

6.2   Textual Features Extraction

The procedure used for extracting textual features information is based on two processing tools, *WebDownload* [14] and *KX* [15]. WebDownload is a preprocessing tool used to download and parse websites, whereas KX, is the main text processing tool.

WebDownload is a robust Java-based tool that crawls, downloads and manipulates the content of HTML pages. It relies on several existing libraries for handling different data formats available through the Web, from HTML pages (based on the Jericho HTML Parser[2]) to XML (based on JDOM[3]) and JSON (based on JSON-SIMPLE[4]).

KX is an unsupervised tool for key-concept extraction that, given a document, identifies a list of relevant concepts and ranks them by assigning a relevance score. It currently supports English, Italian, French, Finnish and Swedish documents.

The process for key-concept extraction performed by KX is displayed in Figure 4. The input data correspond to the web-page content extracted and cleaned by *WebDownload* (the structure information is preserved), while the output is a list of key-concepts ranked by relevance. The first step in the workflow is *tokenisation*, i.e. the segmentation of the text string into words. Then, all possible n-grams composed by any token sequence are extracted, for instance 'condition experts, 'and tornadoes'. In this work, the max length of the selected n-grams is set to four.

Then, from the n-gram list a sublist of multiword expressions (MWE) is derived, i.e. combinations of words expressing a unitary concept, e.g. 'weather service' or 'seven day forecast'. The n-grams are analysed with the morphological analyser of the TextPro toolsuite [16] in order to select as multiword expressions only those n-grams that match certain lexical patterns (i.e. part-of-speech). This is the so-called *linguistic filter*. For example, one of the patterns admitted for 4-grams is the following:

$$[N]\sim[O]\sim[ASPGLU]\sim[NU]$$

This means that a 4-gram is a candidate multiword term if it is composed by a Noun, followed by "of" or "for" (defined as $[O]$), followed by either an Adjective (A), Singular noun (S), Past participle (P), Gerund (G), punctuation (L) or Unknown word (U), followed by either a Noun (N) or Unknown word. This is matched for example by the 4-gram "increase $[S]$ of $[O]$ average $[S]$ temperature $[N]$".

In this step, also a frequency-based threshold is set, so that an n-gram is admissible as MWE only if it occurs at least 2 times in the current document.

In the next step, the single words and the MWEs are ranked by frequency in order to obtain a first list of key-concepts. Thus, frequency is the baseline ranking parameter, based on the assumption that important concepts are mentioned more frequently than less important ones. Frequency is normalized by dividing the number of key-concept occurrences by the total number of tokens in the current document.

Note that the first key-concept list is obtained by applying black and white lists almost at every step of the process, as shown in Figure 5. A black list is applied for

---

[2] http://jericho.htmlparser.net/
[3] http://www.jdom.org/
[4] http://code.google.com/p/json-simple/

discarding n-grams containing one of the language-specific stopwords. Also single words corresponding to stopwords such as 'like', 'many', 'often', are discarded.

A black list is applied also when selecting MWEs from n-grams, in order to eliminate multiwords that may be selected by mistake, such as 'following description, 'in fact', etc. Finally, black and white lists are applied also to the key-concept list, in order to define expressions that should never be selected as relevant key-concepts as well as terms that should always be included in the key-concept rank.

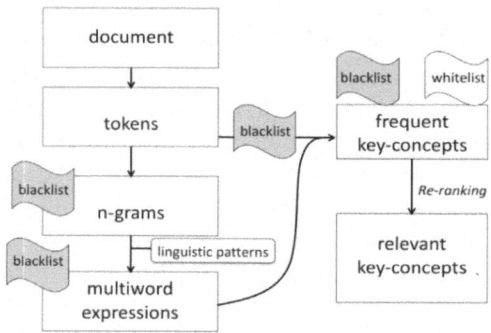

Fig. 5. Key-concept extraction workflow

After a first frequency-based ranking, an additional re-ranking step is performed in order to increase the rank of specific key-concepts based on the following criteria:

1. The frequency score is multiplied by the length in tokens of the key-concept, so that longer key-concepts are not penalized over shorter ones.
2. The frequency score is multiplied by a position factor computed as:

$$PositionFactor = \left(\frac{1 + MaxIndex - KCPosition}{MaxIndex}\right)^2$$

where *MaxIndex* is the length of the current document and *KCPosition* is the position of the first key-concept occurrence in the text. The outcome of this equation should be $0 < PositionFactor \leq 1$. This operation is performed in order to give higher relevance to concepts that are mentioned first in the web site, following the intuition that important concepts are mentioned before less relevant ones. In addition, to take into account the structure of the web page, we boost in a similar way the relevance of key-concepts appearing in headlines or in bold font. Our strategy is to compute for these concepts a position factor as if they appeared at the beginning of the document, by setting *KCPosition = 1*. In this way, we obtain *PositionFactor = 1*, which is the maximum possible factor assigned to key-concepts.

3. The frequency score of nested key-concepts (i.e. appearing within other longer key-concepts) is penalized in two ways: if a key-concept is (stringwise) included in a longer key-concept with a higher frequency-based score, the score of the shorter key-concept is transferred to the count of the longer one. If a key-concept is (stringwise) included in a longer key-concept with a lower relevance, the longer key-concept is assigned the frequency of the shorter one.

At the end of the re-ranking process, the system produces a final list of key-concepts, in which each entry has been associated with a relevance score. Based on these concepts, the selection of the feature vector and of the training/testing datasets is described in the experiments section.

7   Experimental Study

A set of experiments was conducted to evaluate the effectiveness of the "web search" and classification steps, as well as the final results of the system. In this section, we describe the implementation details of these steps, the datasets used and the produced results. Specifically, the presented experiment focuses on retrieving weather providers for the sub-region of Uusimaa, located in the southern part of Finland. However, it is clear that the procedure can be applied to any other region of interest without affecting the other parts of the framework. The selection of the cities was based exclusively on population criteria and they are the following: Helsinki, Espoo, Vantaa, Porvoo, Hyvinkää, Lohja, Järvenpää, Tuusula, Kirkkonummi, Kerava, Vihti and Sipoo.

7.1   Web Search Step

In order to select the most appropriate web search engine for our experiments, we first evaluated existing APIs including Google, Yahoo and Bing. This assessment showed that Google API restricts its use inside a site, while Bing has quite strict terms of use that do not allow processing of the results. However, such limitations were not imposed by Yahoo! BOSS API[5] and therefore the latter was selected as the general-purpose web search engine for the web search step. During this step, both basic and extended queries are generated, the results of which are evaluated based on their precision and recall. However, since the recall measure cannot be computed, given the fact that the number of weather relevant sites in the web is unknown, we consider the number of relevant retrieved websites as a recall indication for our methods. It should be noted that the actual objective of this step is to achieve a high recall, while the improvement of the precision is performed in the classification process.

Basic queries, as mentioned in section 5.1, are created by combining geographical and weather information (7 queries per city). Table 1 contains the queries created for Helsinki city. For the geographic area we target, 84 queries were generated in total.

Table 1. The examples of basic queries submitted for Helsinki

weather + Helsinki	pressure + Helsinki
temperature + Helsinki	rain + Helsinki
"sky condition" + Helsinki	humidity + Helsinki
wind + Helsinki	

---

[5] Yahoo! BOSS API: http://developer.yahoo.com/search/boss/

For each query, the maximum number of sites retrieved from the general-purpose engine was experimentally set to 50 sites. The total number of sites retrieved was 3516. Each one of these sites was manually annotated as weather relevant or not. However, it should be noted that several queries returned common results, which resulted in having a total of 2380 unique sites, from which only 681 were annotated as relevant. Thus, the precision for the simple queries technique was 28.6%.

The generation of the extended queries based on keyword spices required a machine learning procedure. During this procedure, 17 weather related keywords (e.g. barometer, rain, temperature) were sent to a general-purpose search engine and 100 sites were retrieved for each. For cross validation purposes, four variations of a manually annotated data set were created by randomly splitting the web pages into the training and validation sets. Finally, as far as the decision tree is concerned, we employed Id3 algorithm from WEKA [17] tool. The performance of the different variations using the can be found in Table 2. We cite the formulas for Precision of T class and the TP Rate of T class. The metrics for the F class are calculated analogously.

$$Precision\ T = \frac{TP}{TP+FP} \text{ and } TP\ Rate\ T = \frac{TP}{TP+FN}$$

Table 2. Evaluation measures of the decision tree algorithm for different dataset variations

Dataset Variations	Precision of T class	Precision of F class	TP Rate of T class	TP Rate of F class
1st variation	0.508	0.929	0.584	0.906
2nd variation	0.532	0.93	0.584	0.915
3rd variation	0.538	0.926	0.558	0.921
4th variation	0.413	0.915	0.504	0.881
Average	0.498	0.925	0.557	0.906

Based on the results, we conclude that apart from the 4th variation, the others demonstrate similar performance. Thus, we select for our tests the results of the 3rd variation, which demonstrates slightly better precision for True class. The keyword spices produced in this case are: "forecast –instrument –part –surface –comment -climate".

Since, the extended queries are created by adding the keyword spices to the initial/ basic query, an additional 7 queries where generated for each city, creating a total of 84 extended queries. Finally, for each query the maximum number of pages retrieved is 50. Table 3 contains some examples of extended queries for Helsinki. The total number of web pages retrieved is 2571. The number of unique sites is 1061 and the number of sites annotated as relevant is 710. Thus, the precision equals to 66.9%

Table 3. Examples of extended queries for Helsinki

weather + forecast –instrument –part –surface –comment -climate +Helsinki
temperature + forecast –instrument –part –surface –comment -climate +Helsinki
"sky condition" + forecast –instrument –part –surface –comment -climate + Helsinki

Based on the values of precision and retrieved results, we can easily conclude that the keyword spices method demonstrate a better performance, a fact that could question the usage of basic queries. However, although retrieving relevant results is of main interest, it is also important to achieve a high recall. Thus, we need to investigate whether these methods provide the same set of results or they are complementary. Table 4 shows that although the aforementioned techniques return a considerable number of common sites, they also retrieve additional links. Specifically an overlap of 50.8% in average is reported. Therefore, in order to discover as many as possible different weather related sites, we should employ both query techniques.

Table 4. Number of links extracted by the different query techniques

Type of query technique	Number of relevant pages	Number of web pages found only by this technique (i.e. not common web pages)	Common web pages
Basic query	681	328	353
Extended query	710	357	
Both techniques	1391	1038	

7.2   Classification Step

The classification step includes model building with manually annotated training sets and testing with the results from "web search" step.

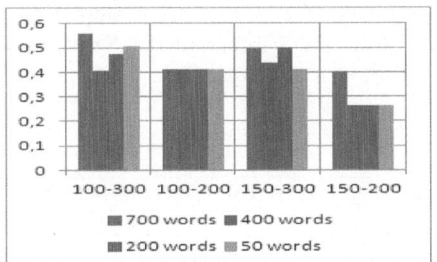

Fig. 6. Accuracy values for different lexicons   **Fig. 7.** F-score values for different lexicons

During training we employed a manually annotated dataset of 650 websites. In order to generate the optimum feature vector (i.e. lexicon) and training set size we have generated several models by splitting the aforementioned dataset into training and validation sets. Specifically, we created a series of different models based on the lexicons and the training data used. The lexicons' size varies from 50 to 700 words, the number of the positive training sites from 100 to 150 and that of negative sites from 200 to 300 sites, while the testing set was 300 websites with 40% of them being relevant. For cross-validation purposes, we created five different variations for each of the aforementioned cases by splitting randomly the training and validation set. We also applied 5-fold cross validation on the training part of the dataset to obtain the best cost and radial function parameters of the model. These models were tested against the validation part of the aforementioned dataset and compared based on the classifier accuracy and F-score values. Figures 6 and 7 depict the values of accuracy and F-score for different models. Based on both figures, we can deduce performance improves when larger lexicons are considered. We also observe that the bigger is the ratio of positive to negative sites, the better the model performs.

Table 5. Precision and Recall for different training set variations

Number of positive – negative sites for training	700 words		400 words		200 words		50 words	
	Precision	Recall	Precision	Recall	Precision	Recall	Precision	Recall
100 - 200	0.411	1	0.410	1	0.410	1	0.410	1
100 - 300	0.619	0.809	0.407	1	0.506	0.884	0.562	0.681
150 – 200	0.367	0.972	0.261	1	0.261	1	0.261	1
150 - 300	0.497	0.951	0.422	0.997	0.472	0.964	0.411	1

Table 5 contains the precision, recall, F-score and accuracy average values (i.e. averaging the results of the 5-fold cross validation) of all the models created. The mean values of precision and recall for all cases are 0.418 and 0.953, while their variance is relatively low, calculated as 0.011 and 0.008 respectively.

In the testing phase, based on the findings during training, we select the model with feature vector size equal to 700 words and with number of the positive- negative training sites 100-300 against the sites returned during the "web search" step. Table 6 contains the precision and recall values, as well as the F-score of the SVM model applied on the sites retrieved from the basic and the extended queries.

Table 6. Precision, Recall and F-score for best SVMs

Lexicon size	Positive – negative sites for training	Basic queries testing results			Extended queries testing results		
		Precision	Recall	F-score	Precision	Recall	F-score
700 words	100-300	0.671	0.789	0.725	0.779	0.811	0.794

7.3   Final Results

In Table 7, we present an overview of the results of the web search and classification steps, as well as the total performance of the proposed framework. As it is observed from Table 4, the total number of web pages retrieved for both types of queries rises to 2992 unique web pages, with only 994 of these pages being actually relevant, which corresponds to less than half of the web pages retrieved. We notice that the use of keyword spices improved the precision of the retrieved results, since it is almost 70% compared to the 29% precision of the basic queries. Finally, from the last two columns of the table, we can easily deduce that the use of SVM improved significantly the results returned by Yahoo! BOSS API, since the precision in the case of basic queries was increased by a 38.5% while the improvement in the case of the extended queries was 11%.

As far as the time performance is concerned, we should note that training step of classification is an offline procedure and realized only once. On the other hand, the testing phase takes place at runtime and takes about 4 seconds for 100 websites.

Table 7. Precision before and after the post-processing step

Testing Set	webpages retrieved by Yahoo!	Retrieved relevant sites	Precision at web-search	Classifi-cation Precision	Classifi-cation Recall	Final retrieved results
Basic queries testing results	2380	681	0.286	0.671	0.789	537
Extended queries testing results	1071	710	0.669	0.779	0.811	576
Total unique results	2992	994	0.332	0.867	0.426	424

8   Conclusions

In this paper, we propose a framework for discovering environmental nodes, which combines state-of-art techniques of domain-specific search and advanced key-phrases extraction tools under a supervised framework. From the experimental results and the comparisons we have performed, it is shown that the combination of the techniques under the classification framework demonstrates better performance than direct employment of these techniques. The current domain-specific discovery framework, although it is evaluated in the environmental domain, could be applied in a variance of areas (e.g. health, cooking) due to the domain independent techniques applied.

Although the selection of Finland and the focus on specific types of weather sites (i.e. sites with measurements) was dictated by the needs of the ongoing European FP7 project PESCaDO[6], aiming at the orchestration of web environmental services to enable personalised decision support and provide tailored environmental information, the same techniques could be employed to deal with other geographical areas and could be applied to the rest of environmental aspects such as air quality and pollen.

Future work includes the use of a more effective classification technique by considering additional features that describe image data, extraction of empirical environmental features, as well as discovery of air quality and pollen environmental nodes. In addition, a more extended evaluation is planned in order to investigate how the system performs in the case of low and high quality basic queries.

Acknowledgments. This work was supported by the project PESCaDO (FP7-248594) funded by the EC.

---

[6] https://www.pescado-project.eu/

References

1. Oyama, S., Kokubo, T., Ishida, T., Yamada, T., Kitamura, Y.: Keyword Spices: A New Method for Building Domain-Specific Web Search Engines. In: Proceedings of the 17th International Joint Conferences on Artificial Intelligence, pp. 1457–1463 (2001)
2. Oyama, S., Kokubo, T., Ishida, T.: Domain-Specific Web Search with Keyword Spices. IEEE Transactions on Knowledge and Data Engineering 16, 17–27 (2004)
3. Menemenis, F., Papadopoulos, S., Bratu, B., Waddington, S., Kompatsiaris, Y.: AQUAM: Automatic Query Formulation Architecture for Mobile Applications. In: Proceedings of the 7th International Conference on Mobile and Ubiquitous Multimedia MUM 2008, Umea, Sweden, December 3-5. ACM, New York (2008)
4. Shakes, J., Langheinrich, M., Etzioni, O.: Dynamic reference sifting: a case study in the homepage domain. In: Proceedings of the 6th International World Wide Web Conference (WWW6), pp. 189–200 (1997)
5. Chen, H., Fan, H., Chau, M., Zeng, D.: MetaSpider: Meta-Searching and Categorization on the Web. Journal of the American Society for Information Science and Technology 52(13), 1134–1147 (2001)
6. Luong, H.P., Gauch, S., Wang, Q.: Ontology-Based Focused Crawling. In: Int. Conference on Information, Process, and Knowledge Management, pp. 123–128 (2009)
7. McCallum, A., Nigam, K., Rennie, J., Seymore, K.: A Machine Learning Approach to Building Domain-Specific Search Engines. In: Proceedings of the 16th International Joint Conference on Artificial Intelligence, pp. 662–667 (1999a)
8. McCallum, A., Nigam, K., Rennie, J., Seymore, K.: Building Domain-Specific Search Engines with Machine Learning Techniques. In: Proc. AAAI 1999 Spring Symposium on Intelligent Agents in Cyberspace (1999b)
9. Chakrabarti, S., van den Berg, M., Byron Dom, B.: Focused crawling: a new approach to topic-specific Web resource discovery. Computer Networks: The International Journal of Computer and Telecommunications Networking 31(11-16), 1623–1640 (1999)
10. Tang, T.T., Hawking, D., Craswell, N., Sankaranarayana, R.S.: Focused crawling in depression portal search: A feasibility study. In: Proceedings of the 9th Australasian Document Computing Symposium, Melbourne, Australia, December 13 (2004)
11. Zheng, H.-T., Kanga, B.-Y., Kim, H.-G.: An ontology-based approach to learnable focused crawling. Information Sciences 178(23), 4512–4522 (2008)
12. Boser, B.E., Guyon, I.M., Va, V.N.: A training algorithm for optimal margin classifiers. In: COLT 1992: Proceedings of the Fifth Annual Workshop on Computational Learning Theory, pp. 144–152. ACM Press, New York (1992)
13. Chang, C.-C., Lin, C.-J.: LIBSVM: a library for support vector machines. ACM Transactions on Intelligent Systems and Technology 2, 27:1–27:27 (2011), http://www.csie.ntu.edu.tw/~cjlin/libsvm
14. Girardi, C.: The HLT Web Manager. FBK Technical Report n. 23969 (2011)
15. Pianta, E., Tonelli, S.: KX: A Flexible System for Keyphrase Extraction. In: Proceedings of SemEval 2010, Uppsala, Sweden (2010)
16. Pianta, E., Girardi, C., Zanoli, R.: The TextPro tool suite. In: Proceedings of the 6th Language Resources and Evaluation Conference (LREC 2008), Marrakech, Morocco (2008)
17. Machine Learning Group at University of Waikato: Weka 3: Data Mining Software in Java, http://www.cs.waikato.ac.nz/ml/weka/

Web Searching
with Entity Mining at Query Time

Pavlos Fafalios[1], Ioannis Kitsos[1], Yannis Marketakis[1], Claudio Baldassarre[2],
Michail Salampasis[3], and Yannis Tzitzikas[1]

[1] Institute of Computer Science, FORTH-ICS,
and Computer Science Department, University of Crete, Greece
[2] Food and Agriculture Organization of the United Nations
[3] Vienna Univ. of Technology, Institute of Software Technology
and Interactive Systems
{fafalios,kitsos,marketak,tzitzik}@ics.forth.gr,
Claudio.Baldassarre@fao.org, salampasis@ifs.tuwien.ac.at

Abstract. In this paper we present a method to enrich the classical
web searching with *entity mining* that is performed at *query time*. The
results of entity mining (entities grouped in categories) can complement
the query answers with useful for the user information which can be further exploited in a *faceted* search-like interaction scheme. We show that
the application of entity mining over the *snippets* of the top-hits of the
answers, can be performed at real-time. However mining over the snippets returns less entities than mining over the full contents of the hits,
and for this reason we report comparative results for these two scenarios. In addition, we show how *Linked Data* can be exploited for specify-
ing the entities of interest and for providing further information about
the identified entities, implementing a kind of entity-based integration
of documents and (semantic) data. Finally, we discuss the applicability
of this approach on professional search, specifically for the domains of
fisheries/aquaculture and patents.

1 Introduction

Entity search engines aim at providing the user with entities and relationships
between these entities, instead of providing the user with links to web pages.
Although this is an interesting line of research and there are already various
entity search engines and approaches [6,20,8], according to our opinion, these
approaches/tools are still in their infancy in the sense that they are not really
useful for the usual information needs of the users. For this reason, instead of
radically changing the way users search for information, we propose *enriching*
the classical interaction scheme of web search engines with entity mining, as a
means to combine the pros of both families of systems. For instance, showing
to the user the recognized categories and entities (e.g. at a left bar of the user
interface), can be useful for the user in various contexts and for various information seeking tasks. For example, consider the case of a user wanting to find

M. Salampasis and B. Larsen (Eds.): IRFC 2012, LNCS 7356, pp. 73–88, 2012.
© Springer-Verlag Berlin Heidelberg 2012

persons who work on a research topic, places related to one particular painter, telephones of restaurants on a particular area, etc. If such entities are made available to the user then he can simply collectively use these entities instead of having to open several pages and searching in each one of them. Furthermore, these entities allow the user to restrict his focus on the part of the answer where a particular entity has been identified. In addition, when the user views an entity that he already knows that is relevant to his information need, this allows him to realize that his query is appropriate for that need. Recognizing categories and entities is not only useful to public web search, but it could be particularly useful in professional search that is, search in the workplace, especially in industrial research and development [14]. For example in professional patent search, in many situations, one must look beyond keywords to find and analyse patents based on a more sophisticated understanding of the patents content and meaning [13]. Technologies such as entity identification and analysis could become a significant aid to such searches and can be seen, together with other text analysis technologies, as becoming the cutting edge of information retrieval science [2].

From an information integration point of view we could say that *entity names* are used as the "glue" for automatically connecting documents with data (and knowledge). This approach does not require deciding or designing an integrated schema/view, nor mappings between concepts as in knowledge bases, or mappings in the form of queries as in the case of databases. The key point is that entities can be identified in documents, data, database cells, metadata attributes and knowledge bases.

To enrich web searching with *Named Entity Mining* (for short *NEM*), we have to tackle (at least) two main challenges: (a) *Real-time Response*: Real time interaction is very important in web searching (and generally in any setting of an interactive search system), however *NEM* is in general computationally expensive in the sense that the required processing time (for extracting entities) is proportional to the size (contents) of the documents. (b) *Selection/Ranking of Entities:* We have to specify criteria that determine the selection and ranking of the (often numerous) discovered entities.

In this work we focus on *NEM* that is performed at *query time* and no preprocessing or indexing has been done. Figure 1 shows an indicative screendump of a prototype system that we have designed and developed[1]. This prototype retrieves the top-K hits (the user is able to set the value of K) of a WSE (Web Search Engine), in our case Google, and mines entities at that time either from the *snippets*, or from the *full contents* (depending on what the user wants) of the top hits of the query answer. The discovered entities are grouped according to their categories and are visualized and exploited according to the *faceted exploration* interaction paradigm [18]: when the user clicks on an entity, the hits are restricted to those that contain that entity, and so on. Furthermore, the system, after user's request, can apply mining over a desired hit and discover all entities of that hit. All these are performed at query time, without any preprocessing.

---

[1] Accessible through http://www.ics.forth.gr/isl/ios

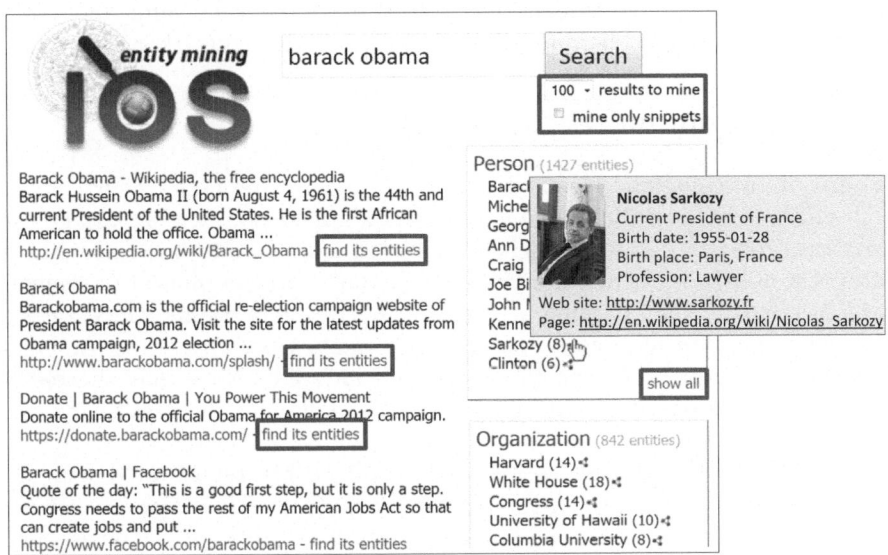

Fig. 1. A prototype offering Web Searching enriched with entity mining

In Figure 1 we can see what the user is getting after having submitted the query "Barack Obama" and selected to mine the full contents of the top-100 results. We observe entities like *Joe Biden, John McCain, Michelle Obama* and *Clinton* under the category "Person", and *White House, Congress, Columbia University* and *Harvard* under the category "Organization". Only the top-10 entities of each category are shown, but the user can see all of them by clicking on "show all". By clicking on an entity, e.g on *Joe Biden*, the search space narrows to those containing that entity, while with "find its entities" on a hit, the user can ask and get all entities of that hit. Furthermore, for each entity the user can ask the system to fetch more information from the *Linked Open Data* cloud (e.g. as we see for Nicolas Sarkozy in Figure 1).

In a nutshell, in this paper: (a) we detail a novel combination of *NEM* technologies for *enriching* the classical web (meta) searching process with entity mining performed at query time, where the mined entities are exploited for offering faceted exploration, (b) we compare the results of *NEM over document snippets* versus *NEM* over the *full document contents* according to various perspectives (mined entities, computational cost), (c) we elaborate on the *ranking of entities* and we report the results of a comparative evaluation with users, and (d) we show how to exploit the LOD (Linked Open Data) cloud for enriching the identified entities with links to their corresponding semantic descriptions.

The rest of this paper is organized as follows. Section 2 discusses the context, related works, and possible approaches for enriching web and professional search with *NEM*. Section 3 describes the approach that we investigate in this work (architecture, entity ranking, LOD-based enrichment), and a prototype over the fisheries/aquaculture domain. Section 4 reports experimental results, and finally, Section 5 concludes and identifies directions that are worth further research.

2   Background and Related Work

The idea of enriching the classical *query-and-response* process of current web search engines, with *static* and *dynamic* metadata for supporting *exploratory search* was proposed in [17] and it is described in more detail (enriched with the results of a user-based evaluation) in [16]. In that work the notion of *dynamic metadata* refers to the outcome of *results clustering* algorithms which take as input the *snippets* of hits, where snippets are *query word dependent* (and thus they cannot a-priori be extracted, stored and indexed). Note that the results of *NEM* if applied over the textual snippets also falls into the case of *dynamic metadata*.

We can model our setting as follows. Let D be the set of all documents, and C the set of all supported categories, e.g. $C = \{Locations,\ Persons,\ Organizations,\ Events\}$. Let M be the set of all minable entities where each entity is described by a string. We can now model the functionality of a *NEM* tool by two functions mc and mec. Specifically, let $mc : D \to 2^C$ be the function that takes as input a document (say $d1$) and returns the categories of the entities that have been recognized in $d1$, e.g. $mc(d1) = \{Locations, Persons\}$. Now let $mec : D \times C \to 2^M$ be the function that takes as input a document d and a category c and returns entities belonging to that category which have been recognized in that document. For example, $mec(d1, Location) = \{Crete, Athens\}$.

There are several approaches that could be used in order to enrich the classical web searching with *NEM*. Some of them are described below.

$\mathcal{RS}$: **Real-time *NEM* over the __S__nippets of the top hits** of the answer. Here entity mining is performed only over the *snippets* of the top hits of the returned answer.

$\mathcal{RC}$: **Real-time *NEM* over the __C__ontents of the top hits** of the answer. Here the full contents of the top hits of the returned answer are downloaded and then entity mining is performed. Clearly, this process can take much more time than $\mathcal{RS}$.

$\mathcal{OC}$: **Off-line *NEM* over the entire __C__orpus**. Here we mine all entities of the corpus *offline* (assuming that the corpus is available), and we build an appropriate index (or database) for using it at run time. For each incoming query, the entities of the top-K (e.g. $K = 100$) hits of the answer are fetched from the index, and are given to the user. An important observation is that the size of the entity index in the worst case could be in the scale of the corpus. Also note that this approach cannot be applied in an uncooperative search environment where full access to the resources is not given.

$\mathcal{OFQ}$: **Offline *NEM* over the top hits** of the answers of the **__F__requent Queries**. Here, also *offline*, for each frequent query of the log file (e.g. for those which are used for query suggestion), we compute its answer, we fetch the top-K hits, we apply *NEM* and save its results as they should be shown (i.e. what the left bar should show) using the approach and indexes described at [10,9]. The benefit of this approach in comparison to $\mathcal{OC}$ is that we do not have to apply *NEM* at the entire collection but only at the top hits of the most frequent queries. This significantly reduces the required computational effort and storage space. The downside of this approach is that if a user submits a query which does not belong to the frequent queries, and thus it has not been processed, then the system cannot provide and suggest entities. In that case the system could offer the choice to the user to apply *NEM* at query time, i.e. approach $\mathcal{RS}$ or $\mathcal{RC}$ as described earlier. Finally, we should note that this approach is applicable also at a meta search level since it does not require access to the entire corpus (only to the query answers), but periodically the index has to be refreshed (mainly incrementally).

There is a plethora of related works and systems that offer a kind of *entity search*, below we briefly describe few of them.

The *Entity Search Engine* [6,5,4] supports only two categories (phone and email) and users have to type formatted queries (using # to denote entities). *NEM* is applied over the entire corpus and the extracted entities are stored in the form of ordered lists based on document ID (much like storing inverted indices), in order to provide their results instantly.

EntityCube[2] is an entity search engine by Mircosoft which extracts entities and relationships from semistructured as well as natural-language Web sources. The goal is to automatically construct and maintain a knowledge base of facts about named entities, their semantic classes, and their mutual relations as well as temporal contexts.

MediaFaces [20,21] provides faceted exploration of media collections and offers a *machine learned* ranking of entity facets based on user click feedback and features extracted from three different ranking sources. For a given entity of interest, they have collected (from knowledge bases like Wikipedia and GeoPlanet) a large pool of related candidate facets (actually related entities).

The approach described at [8] aims at identifying related entities by analyzing a user query (that describes one entity in a TREC-like manner) and then generating and sending (to various search engines) an enriched query, and finally analyzing the (full contents) of the returned results.

With respect to the approaches $\mathcal{RS}$, $\mathcal{RC}$, $\mathcal{OC}$ and $\mathcal{OFQ}$ described earlier, most systems follow approach $\mathcal{OC}$. [8] follows the $\mathcal{RC}$ approach, while the only system that offers $\mathcal{OFQ}$ is [9]. To the best of our knowledge the current paper is the first that investigates the $\mathcal{RS}$ approach.

3   *NEM* at Query Time

We focus on a *dynamic* approach where no pre-processing of the resources has been done. Analogously to works like [22] which compares the outcomes of *clustering over snippets* with the outcomes of *clustering over contents*, in this work

---

[2] http://entitycube.research.microsoft.com/

we investigate the same question but for the case of *entity mining*. Furthermore we investigate linking the identified entities with *Linked Data*.

Architecture. Our prototype system IOS, supports the approaches $\mathcal{RS}$, $\mathcal{RC}$ and $\mathcal{OFQ}$. The default choice is $\mathcal{RS}$ and $\mathcal{OFQ}$, while $\mathcal{RC}$ is offered on demand (see the options at the upper right corner of Figure 1). Approach $\mathcal{OFQ}$ is offered for the frequent queries only and the reason for providing it is because: (i) it can offer instant behavior for a significant percentage of incoming queries, (ii) it is less space consuming than the $\mathcal{OC}$ approach that mines everything, and (iii) it is beneficial for the server since it reduces the number of incoming queries and the same precomputed information is exploited in several requests. We do not further analyze this case since it has been described in detail at [10] and [9]. Figure 2 shows the architecture. We currently use *GateAnnie*[3][3,7] for *NEM*. In our case it takes as input a set of documents (or document snippets), specifically those of the top-K hits of the query answer. It returns as output a set of lists (one list for each category). In general, *GateAnnie* relies on finite state algorithms and the JAPE (regular expressions over annotations) language. It consists of various components, in our case the following are used: *Unicode Tokeniser* (for splitting text into simple tokens such as numbers, punctuation and words), *Gazetteer* (predefined lists of entity names), and *Sentence Splitter* (for segmenting text into sentences). The prototype also supports *faceted search-like* restriction of the answer, i.e. the user is able to *gradually* select entities from one or more categories and refine the answer set accordingly (the mechanism is *session*-based). So far all such selections have disjunctive (OR) semantics.

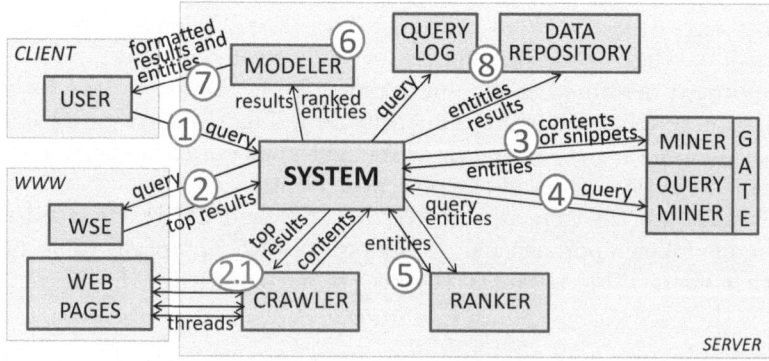

Fig. 2. The architecture (and main flow of control) of the prototype system

Entity Ranking. Entity selection and ranking is important since usually the UI has limited space therefore only a few can be shown at the beginning. We propose tackling this problem by (a) ranking all identified entities for deciding

---

[3] http://gate.ac.uk/ie/annie.html

the top-10 entities to be shown for each category, and (b) offer to user the ability to show more entities (all) on demand. Below we focus on ranking methods that do not rely on any log analysis, so they are aligned with the dynamic nature of our approach.

Consider a query q and let A be the set of returned hits (or the top-K hits of the answer). Now consider a category c. The entities that fall in this category are $E = \cup_{a \in A} mec(a, c)$. For an $e \in E$, let $docs(e) = \{a \in A \mid e \in mec(a, c)\}$, i.e. $docs(e)$ is the set of documents in which entity e is identified.

We need a method to rank the elements of E. One approach is to count the elements of A in which the entity appears, i.e. its *frequency*. Furthermore, we can take into account the rank of the documents that contain that entity in order to promote those entities that are identified in more highly ranked documents (otherwise an entity occurring in the first two hits will receive the same score as one occurring in the last two). For an $a \in A$, let $rank(a)$ be its position in the answer (the first hit has rank equal to 1, the second 2, and so on). We can capture this requirement by a formula of the form:

$$Score_{rank}(e) = \sum_{a \in docs(e)} ((|A| + 1) - rank(a)) \tag{1}$$

We can see that an occurrence of e in the first hit, counts $|A|$, while an occurrence in the last document of the answer counts for 1.

Another approach is to take into account the *words* of the entity name and the query string. If for an entity $e \in E$, we denote by $w(e)$ the words of its name, and by $w(q)$ the words of the query, then we can define $Score_{name}(q, e) = \frac{|w(q) \cap w(e)|}{|w(e)|}$. To tolerate small differences (due to typos or lexical variations), we can define an alternative scoring function that is based on the *Edit Distance*:

$$Score_{nameDist}(q, e) = \frac{|\{ a \in w(q) \mid \exists b \in w(e), EDist(a, b) \le 2\}|}{|w(q)|} \tag{2}$$

which returns the percentage of the words of q which can be "matched" to one word of the entity e either exactly or up to an Edit distance equal to 2.

The above scores can be combined to reach a final score that considers both perspectives. We can adopt the *harmonic mean* for promoting those entities which have high scores in both perspectives. However notice that if an entity has not any query word (or a word that is close to a query word), that entity would take zero at $Score_{name}$ and that would zero also the harmonic mean. One approach to tackle this problem is to compute the plain (instead of the harmonic) mean, or in place of $Score_{nameDist}(q, e)$ to have the sum $Score_{nameDist}(q, e) + b$ for a very small positive constant b (e.g. $b = 0.01$). To conclude we could use:

$$Score(q, e) = \frac{2 \; Score_{rank}(q, e) \; Score_{nameDist}(q, e)}{Score_{rank}(q, e) + Score_{nameDist}(q, e)} \tag{3}$$

3.1   On Exploiting Linked Open Data

There are already vast amounts of structured information published according to the principles of *Linked Open Data (LOD)*. The availability of such datasets enables not only to configure easily the entity names that are interesting for the application at hand, but also the enrichment of the identified entities with more information about them. In this way the user not only can get useful information about one entity without having to submit a new query, but he can also start browsing the entities that are linked to that entity.

Another important point is that exploiting LOD is more dynamic, affordable and feasible, than an approach that requires each search system to keep stored and maintain its own knowledge base of entities and facts. Returning to our setting, the first question is which LOD dataset(s) to use. One approach is to identify and specify one or more appropriate datasets for each category of entities. For example, for entities in category "Location", the *GeoNames*[4] dataset is ideal since it offers access to millions of placenames. Furthermore, *DBpedia*[5] is appropriate for multiple categories such as "Organizations", "People" and "Locations". Other sources that could be used include: *FreeBase*[6] *(for persons, places and things)*, YAGO[19] *(for Wikipedia, WordNet and GeoNames)*, Wiki-company[7] (for organizations). In addition *FactForge*[1] includes 8 LOD datasets (including DBpedia, Freebase, Geonames, UMBEL, Wordnet). DBpedia and FactForge offer access through SPARQL endpoints[8].

Running one (SPARQL) query for each entity would be a very expensive task, especially if the system has discovered a lot of entities. Some possible ways to tackle this problem are: (a) offer this service on demand, (b) for the frequent queries pay this cost at pre-processing time and exploit the results as described in [10,9], (c) periodically retrieve and store locally all entities of each category, so at real time only a matching process is required (however here we have increased space and maintenance requirements). Note however that approach (c) is essentially the approach of our prototype (even though no Linked Data are used), since the Gazetteers of *GateAnnie* that we use include names of persons (11,974), organizations (8,544), and locations (29,984); in total about 50,502 names are used in our setting. Furthermore, lists of prefixes and postfixes are contained that aid the identification of entities (e.g. from a phrase "Web *inventor* Tim Berners-Lee", it recognizes "Tim Berners-Lee" as a *person* due to the prefix "inventor"). So the essential difference could be the following: instead of having a *NEM* component that contains predefined named lists/rules, it is beneficial (certainly from an architectural point of view) to offer the ability to the system to download the required lists (from the constantly evolving LOD) that are appropriate for the application at hand. For example, we can run a SPARQL query that returns a list with all objects of `rdf:type dbp-ont:Artist` and thereby offer the ability to explore artists in the search results.

---

[4] `http://www.geonames.org/`
[5] `http://dbpedia.org/`
[6] `http://www.freebase.com/`
[7] `http://wikicompany.org/`
[8] DBpedia: `http://dbpedia.org/sparql`, FactForge: `http://www.factforge.net/sparql`

Currently our prototype adopts the (a) approach for the general web search scenario, and the (c) approach for vertical search scenarios. Specifically when the user clicks on the small icon at the right of an entity, the system at that time checks if that entity lies in the LOD cloud (by performing a SPARQL query) and if yes it collects more information about that entity which are visualized in a popup window as shown in Figure 1. For instance, the following query, evaluated over *FactForge* SPARQL Endpoint, returns the basic information about a *foaf:Person* with *foaf:name 'Barack Obama'*:

```
SELECT DISTINCT
?person ?name ?comment ?birthDate ?birthPlace ?homepage ?thumbnail WHERE {
?person rdf:type foaf:Person; foaf:name ?name
FILTER(regex(str(?name),'Barack Obama','i'))
?person dbp-ont:birthDate ?birthDate .
OPTIONAL{?person rdfs:comment ?comment
          FILTER(langMatches(lang(?comment),"EN"))}
OPTIONAL{?person dbp-ont:thumbnail ?thumbnail}
OPTIONAL{?person foaf:homepage ?homepage}
OPTIONAL{?person dbp-ont:birthPlace ?place .
?place rdfs:label ?birthPlace FILTER(langMatches(lang(?birthPlace), "EN"))}
```

Case Study: Fisheries and Aquaculture Publications. Apart from the case of general purpose Web searching, we have started investigating this approach in vertical search scenarios. One of this is the domain of *FAO* (Food and Agriculture Organization) publications about *fisheries and aquaculture*. The underlying keyword search system is the *FIGIS search component*[9] which can receive queries through an HTTP API. The search result apart from formatted HTML can be returned in XML format which uses Dublin Core schema to encapsulate bibliographic information. Each returned hit has various textual elements, including publication title and abstract. The first is around 9 words, the second cannot go beyond 3,000 characters. As concern NEM, we identified the following relevant categories: *Countries, Water Areas, Regional Fisheries Bodies*, and *Marine Species*. For each one there is a list of entities: 240 countries, 28 water areas, 47 regional fisheries bodies and 8,277 marine species, in total 8,592 names. Each such entity is also described and mutually networked in the *Fisheries Linked Open Data (FLOD)* RDF dataset. FLOD extended network of entities is exposed via a public SPARQL endpoint[10] and web services.

The objective is to investigate how to enrich keyword search with entity mining where the identified entities are linked to entities in FLOD endpoint, and from which semantic description can be created and served. A screendump of this prototype is shown in Figure 3. The link in the popup window redirects the users to the FLOD graph browser (a customized version of the Pubby web application[11] interfacing with FLOD endpoint).

---

[9] http://www.fao.org/fishery/search/en
[10] http://www.fao.org/figis/flod/endpoint/sparql
[11] http://www4.wiwiss.fu-berlin.de/pubby/

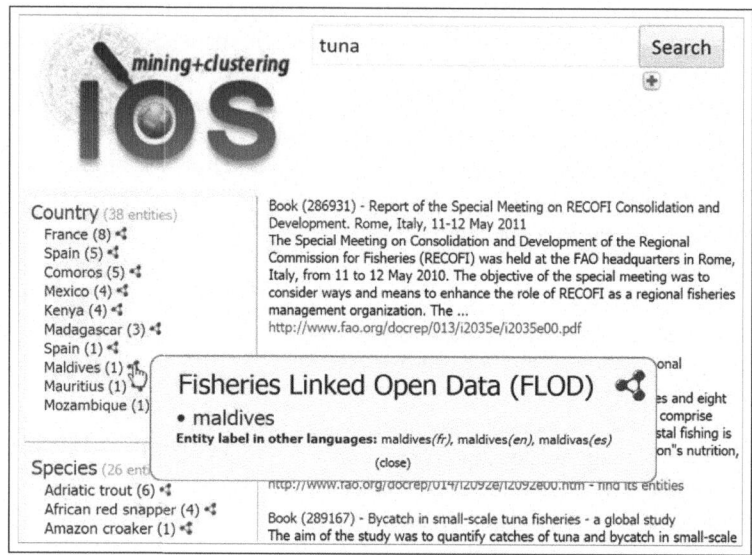

Fig. 3. A prototype over FAO publications with links to FLOD

Patent Search. Entity mining at query time can also be beneficial for patent search. *Patent search* is a kind of professional search, and most patent searches (e.g. patentability and validity) are crucially important for businesses' patent management success. Missing relevant documents is unacceptable therefore the retrieval of all relevant documents is usually necessary. Clearly, this is a kind of *recall-oriented* search, and thus the support of an interactive and gradual (session-based) multi-faceted search process is required. In that context the provision of facets that correspond to various kinds of entities can help the user to get an overview and to quickly restrict the search space. The usefulness of entity mining in patent search is also signified by the emergence of systems like *quantalyze*[12] in which quantities such as temperatures are spotted in the documents (patent documents in this case), their respective semantic context is identified and the quantity itself is normalized to a standard unit. However, more kinds of entities would be useful to be supported, e.g. companies, countries, persons (of related publications), product types, laws, other patents, etc. Since most of them are *named* entities the exploitation of LOD is indispensable. In this setting *NEM* could be applied not only over the textual snippets returned by a simple search, but also over the *abstracts* or *descriptions* (full contents) of the patents.

4   Experimental Results

At section 4.1 we report the results of a comparative evaluation of the three *entity ranking methods* by users, while at section 4.2 we compare *snippet-mining* versus

---

[12] https://www.quantalyze.com/en/

contents-mining from various perspectives. Finally, in section 4.3 we report the results of LOD-related experiments.

4.1   Comparative Evaluation of Entity Ranking Methods by Users

We comparatively evaluated with users the *three* methods for entity ranking, i.e. equations (1), (2) and (3) of Section 3. Fifteen users participated in the evaluation with ages ranging from 20 to 28, 73.3% males and 26.6% females. We selected a set of 20 queries and for each one we printed one page consisting of three columns, one for each ranking method. Each column was showing the top-10 entities for the categories *Person, Location, Organization*. NEM over full contents was used. Each participant was asked to mark the most preferred ranking. If the user could not identify the most preferred, the user could mark two or even all three as equally preferred. We aggregated the results based on plurality ranking (by considering only the most preferred options). The results showed that the most preferred ranking is that derived by equation (1), which was the most preferred in 228 of the 15x20=300 questions. The equations (2) and (3) received almost the same preference; they were selected as most preferred options in 43 and 44 of the 15x20=300 questions.

In more detail, Figure 4a shows that for 13 of the 15 participants, equation (1) was the most preferred, while Figure 4b shows that equation (1) was the most preferred for all 20 queries.

From these we can conclude that the string similarity between the query and the entity name did not improve entity ranking in our setting.

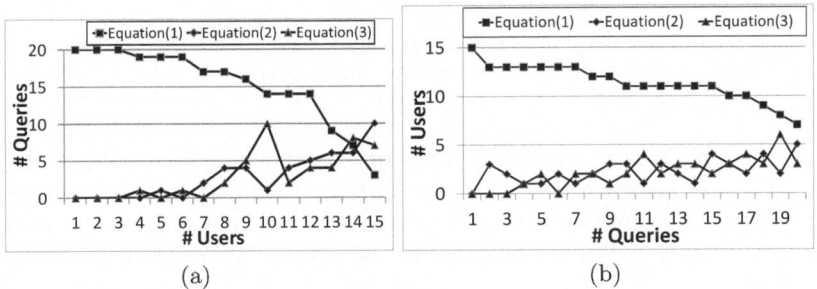

(a)   (b)

Fig. 4. Left: Aggregated preferences for each user. Right: Aggregated preferences for each query.

4.2   Contents Mining versus Snippet Mining

Since a snippet is part of a document, the entities discoverable in a snippet are *subset* of those discoverable at the full document contents. From this perspective we could say that the results of snippet mining are "sound" with respect to the results of documents mining.

To check how different they are we performed various measurements. For a set of 1,000 queries we compared the results of *snippet-mining* and *contents-mining* over

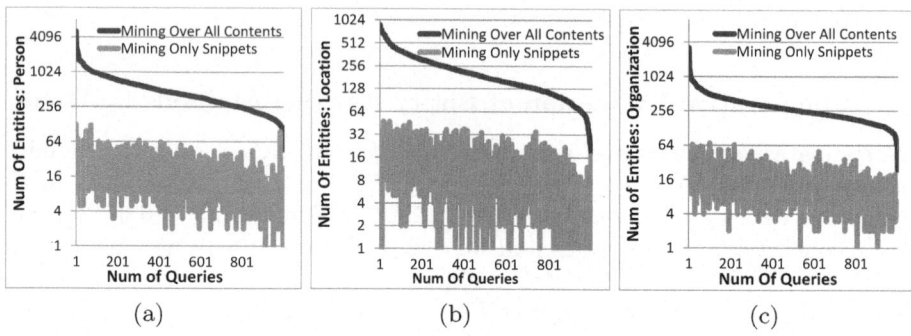

Fig. 5. Comparing the number of mined entities (for the categories *Person, Location, Organization*) over the contents and over the snippets for 1,000 queries (for each query its top-50 hits are mined)

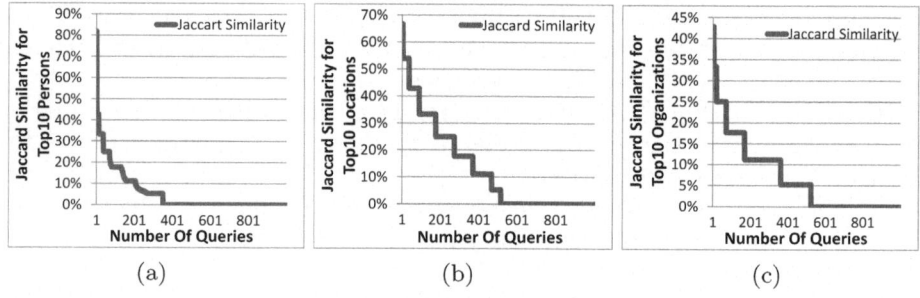

Fig. 6. Jaccard Similarity between top-10 mined entities for the categories *Person, Location, Organization* over snippets and over full contents for 1,000 queries (for each query answer the top-50 hits are considered)

the top-50 hits of the query answers. In our experiments we considered the categories *Person, Location* and *Organization*. The results are shown in Figure 5a, 5b and 5c respectively. The y-axis is in log scale and the queries are ordered in descending order with respect to the number of mined entities over their full contents.

Figure 5a shows that the average number of identified *persons* over full contents is about 527, while the average number of identified persons over snippets is about 18, meaning that content mining yields around 29 times more persons. We should also note that 50% of the queries return less than 500 entities, 43% of queries retrieve from 500 to 1000 entities and only 7% return more than 1000 entities.

In Fig. 5b we observe the same pattern for *locations*: contents-mining in average returns about 219 entities per query while snippet-mining about 12 entities. Finally, according to Fig. 5c, contents mining identifies on average 309 organizations while snippet-mining 14 organizations (22 times less).

To sum up, we could say that contents mining yields around 20 times more entities than snippet mining.

In addition, we compared only the top-10 mined entities as produced by equation (1). We compared these lists as sets using the Jaccard similarity. The results are shown in the three diagrams of Figure 6. We observe that the entities in the category "Person" have Jaccard similarity 0% for the 65% of queries and more than 20% for less than the 10% of queries. For the categories "Location" and "Organization", about half of the queries have 0% similarity. Furthermore, for entities in "Organization" there are no queries with Jaccard similarity more than 50%, while for entities in "Location" there is not any query with more than 70% similarity. From the above we can conclude that the top-10 entities yielded from snippets mining are quite different from those yielded from contents mining. This is quite expectable since (as we described earlier) contents mining yields around 20 times more entities than snippets mining.

Computational and Memory Costs

Below we report execution times. All experiments were carried out using a laptop with processor Intel Core 2 Duo T8300 @ 2.40Ghz CPU, 4GB RAM and running Ubuntu 10.04 (32 bit), and Google was used as the underlying engine. The implementation of the system is in Java 1.6 (J2EE platform), using Apache Tomcat 7 (2GB of RAM).

Snippet-Mining. *Time:* The whole process over the top-50 snippets (each having the size of 193 bytes in average) for one query takes 1.5 seconds in average. The process comprises the following steps. At first we retrieve the results pages from the underlying WSE which costs about 547 ms (36,2% of the total time). Second, we apply *NEM* over the snippets of the retrieved query answers which takes about 901 ms (60,4% of the total time). Third, we apply *NEM* over the query string which costs about 5,6 ms (0,37% of the total time). Finally, we create a string representation of the first page of results with cost about 36 ms (2,4% of the total time) and also a string representation of all entities in about 9 ms (0,6% of total time). The time for ranking entities and categories is negligible (less than 1 ms). In more detail, and only for the *NEM* task, some indicative times follow: 0.2 secs for 10 snippets of total size 0.1 MB, 1.2 secs for 100 snippets of total size 1.94 MB.

The average main memory requirements for one query (for the whole process) is about 37MB.

Contents-Mining. *Time:* The whole process over the top-50 full documents (of total size about 6.8 MB) for one query takes 78 seconds in average. The retrieval of results from the underlying WSE costs less than one second (1% of the total time). The downloading of the contents of each hit costs about 28 seconds (36% of the total time). The application of *NEM* over the contents of the downloaded documents takes about 45 seconds (57% of the total time). The creation of the string representation of the first page of results costs about 33 ms (0.04% of the total time), while the construction of the string representation of all entities takes about 4,5 seconds (6% of total time)[13]. The sorting of the categories and entities

---

[13] Notice that this step takes 6%, while in snippets it takes only 0.6% of the total time. This is due to the much higher number of entities.

costs only a few ms. Some indicative times for *NEM* only: 5.2 seconds for 10 documents of total size 1.5 MB, 107 seconds for 100 documents of total size 16.3 MB.

The average main memory requirements for one query (the whole process) is about 300 MB.

Synopsis. The comparison between snippet and contents mining can be summarized as:

	$\mathcal{RS}$	$\mathcal{RC}$
entities per hit:	1.2	10.1
overall time:	1.5 secs	78 secs
main memory footprint for one query:	37 MB	300 MB

4.3   Linked Data-Related Experimental Results

In general the time required for getting the *semantic description of an entity* by querying a SPARQL endpoint, depends on the particular query and the particular endpoint. In our case, and using the FLOD Endpoint, the average time for an entity of the category *Location* is 230 ms, while the average time for the rest categories is 450 ms.

Below we report the times required for getting the *entire list of entities* of the FAO categories mentioned earlier, by submitting one query for each category: Countries (240) in about 450 ms, Species (8,277) in about 12,000 ms, Water areas (28) in about 300 ms, and Regional Fisheries Bodies (47) in about 340 ms.

5   Concluding Remarks

In this paper we have discussed methods to enhance web searching with entity mining. Such enhancement is useful because it gives the user an overview of the answer space, it allows the user to restrict his focus on the part of the answer where a particular entity has been mined, it is convenient for user needs (or user tasks) that require collecting entities, and it can assist the user to assess whether the submitted query is the right one.

We described four main approaches for supporting this functionality and we focused on two *dynamic methods*, i.e. methods that are performed at query time and do not require any pre-processing. Since such methods have not been studied in the literature (nor supported by existing systems), we compared the application of *NEM* over textual snippets versus *NEM* over the full contents (after having downloaded them at real time) of the top hits (according to various criteria). In brief, the experimental results showed that real time *NEM* over the top snippets is feasible (requires less than 2 secs for the top-50 hits) and yields about 1.2 entities per snippet. On the other hand the approach "download and mine over the full contents" is more time consuming (requires 80 secs for the top-50), but mines much more entities (in average 10.1 per hit).

As regards *entity ranking* we comparatively evaluated three methods; one based on the frequency of the entity and the rank of the hits in which it occurs, one based on similarity with the query string, and one that combines both. The user study showed that the string similarity between the query and the entity name did not improve entity ranking in our setting. Another important point is that the top-10 entities derived from snippet mining and the top-10 entities derived from contents mining for the same queries are quite different; their Jaccard similarity is less than 30% for the majority of the queries. Therefore one issue that is worth further research is to compare the *quality* of the identified entities in snippets versus those identified in contents. Towards the same direction, it is worth investigating approaches for entity deduplication and cleaning that are appropriate for our setting.

In addition, and since there are already vast amounts of structured information published according to the principles of *Linked Open Data (LOD)*, we discussed and showed how they can be exploited for enriching the semantic descriptions of identified entities. In this way, the user not only can get useful information about one entity without having to submit a new query, but he can also start browsing the entities that are linked to that entity.

In future we plan to apply and evaluate empirically this approach in the domains of fisheries/aquaculture and patent search. Regarding the latter, we currently develop a patent search system for the whole spectrum of patent users based on the *ezDL* system[14] (framework for interactive search applications and system for performing evaluations [12]). The plan is to integrate multiple patent data sources, patent search tools and UIs, and one of these tools will offer *NEM*.

The long term vision is to be able to mine not only correct entities but probably entire conceptual models that describe and relate the identified entities (plus other external entities) and are appropriate for the context of the user's information need. After reaching that objective the exploratory process could support the interaction paradigm of faceted search over such (crispy or fuzzy) semantic models, e.g. [11] for plain RDF/S, or [15] for the case Fuzzy RDF.

Acknowledgement. Work done in the context of the *iMarine* (FP7 Research Infrastructures, 2011-2014) and *MUMIA* (COST action IC1002, 2010-2014).

References

1. Bishop, B., Kiryakov, A., Ognyanov, D., Peikov, I., Tashev, Z., Velkov, R.: Factforge: A fast track to the web of data. Semantic Web 2(2), 157–166 (2011)
2. Bonino, D., Ciaramella, A., Corno, F.: Review of the state-of-the-art in patent information and forthcoming evolutions in intelligent patent informatics. World Patent Information 32(1) (2010)
3. Bontcheva, K., Tablan, V., Maynard, D., Cunningham, H.: Evolving GATE to meet new challenges in language engineering. Nat. Lang. Eng. 10, 349–373 (2004)
4. Cheng, T., Chang, K.C.C.: Entity search engine: Towards agile best-effort information integration over the web. In: Proc. of CIDR, pp. 108–113. Citeseer (2007)
5. Cheng, T., Yan, X., Chang, K.C.C.: Entityrank: searching entities directly and holistically. In: Procs. of the 33rd Intern. VLDB Conf., pp. 387–398 (2007)

---

[14] http://www.ezdl.de/

6. Cheng, T., Yan, X., Chang, K.C.C.: Supporting entity search: a large-scale prototype search engine. In: Proceedings of the 2007 ACM SIGMOD International Conference on Management of Data, pp. 1144–1146. ACM (2007)

7. Cunningham, H., Maynard, D., Bontcheva, K., Tablan, V.: GATE: A Framework and Graphical Development Environment for Robust NLP Tools and Applications. In: Procs of the 40th Anniversary Meeting of the Association for Computational Linguistics, ACL 2002 (2002)

8. Ernde, B., Lebel, M., Thiele, C., Hold, A., Naumann, F., Barczyn'ski, W., Brauer, F.: ECIR - a Lightweight Approach for Entity-centric Information Retrieval. In: Proceedings of the 18th Text REtrieval Conference, TREC 2010 (2010)

9. Fafalios, P., Kitsos, I., Tzitzikas, Y.: Scalable, flexible and generic instant overview search. In: WWW 2012 (Demo Paper), Lyon (April 2012)

10. Fafalios, P., Tzitzikas, Y.: Exploiting Available Memory and Disk for Scalable Instant Overview Search. In: Bouguettaya, A., Hauswirth, M., Liu, L. (eds.) WISE 2011. LNCS, vol. 6997, pp. 101–115. Springer, Heidelberg (2011)

11. Ferré, S., Hermann, A.: Semantic Search: Reconciling Expressive Querying and Exploratory Search. In: Aroyo, L., Welty, C., Alani, H., Taylor, J., Bernstein, A., Kagal, L., Noy, N., Blomqvist, E. (eds.) ISWC 2011, Part I. LNCS, vol. 7031, pp. 177–192. Springer, Heidelberg (2011)

12. Fuhr, N.: An infrastructure for supporting the evaluation of interactive information retrieval. In: Procs of the 2011 Workshop on Data Infrastructures for Supporting Information Retrieval Evaluation, DESIRE 2011, NY, USA (2011)

13. Joho, H., Azzopardi, L., Vanderbauwhede, W.: A survey of patent users: an analysis of tasks, behavior, search functionality and system requirements. In: Procs of the 3rd Symposium on Information Interaction in Context. ACM (2010)

14. Kohn, A., Bry, F., Manta, A.: Professional Search: Requirements, Prototype and Preliminary Experience Report (2008)

15. Manolis, N., Tzitzikas, Y.: Interactive Exploration of Fuzzy RDF Knowledge Bases. In: Procs of the 8th Extended Semantic Web Conference, ECSW 2011 (2011)

16. Papadakos, P., Armenatzoglou, N., Kopidaki, S., Tzitzikas, Y.: On exploiting static and dynamically mined metadata for exploratory web searching. Knowledge and Information Systems 30, 493–525 (2012), doi:10.1007/s10115-011-0388-2

17. Papadakos, P., Kopidaki, S., Armenatzoglou, N., Tzitzikas, Y.: Exploratory Web Searching with Dynamic Taxonomies and Results Clustering. In: Agosti, M., Borbinha, J., Kapidakis, S., Papatheodorou, C., Tsakonas, G. (eds.) ECDL 2009. LNCS, vol. 5714, pp. 106–118. Springer, Heidelberg (2009)

18. Sacco, G.M., Tzitzikas, Y.: Dynamic taxonomies and faceted search: theory, practice, and experience, vol. 25. Springer-Verlag New York Inc. (2009)

19. Suchanek, F.M., Kasneci, G., Weikum, G.: Yago: a core of semantic knowledge. In: Procs. of the 16th World Wide Web Conf., pp. 697–706 (2007)

20. van Zwol, R., Garcia Pueyo, L., Muralidharan, M., Sigurbjörnsson, B.: Machine learned ranking of entity facets. In: Procs. of the 33rd Intern. ACM SIGIR Conf., pp. 879–880. ACM (2010)

21. van Zwol, R., Sigurbjornsson, B., Adapala, R., Garcia Pueyo, L., Katiyar, A., Kurapati, K., Muralidharan, M., Muthu, S., Murdock, V., Ng, P., et al.: Faceted exploration of image search results. In: Procs. of the 19th World Wide Web (2010)

22. Zamir, O., Etzioni, O.: Web document clustering: A feasibility demonstration. In: Procs of SIGIR 1998, Melbourne, Australia, pp. 46–54 (1998)

Patent and Norm Exploration
with the m2n Knowledge Discovery Suite

Tobias Schleser[1], Doris Ipsmiller[1], and Roland Mörzinger[2]

[1] m2n – Consulting and Development GmbH, Linz & Vienna, Austria
{schleser,ipsmiller}@m2n.at
[2] Joanneum Research Forschungsgesellschaft mbH, Graz, Austria
roland.moerzinger@joanneum.at

Abstract. Information relevant for patent officers as well as research staff is hidden in huge amounts of patents – both, patents' text and images being valuable knowledge sources. The retrieval process to discover this knowledge is time consuming and therefore costly. Missing the relevant information may even be mission critical. This industry paper presents key functionality of a search and analysis solution built on top of the m2n Knowledge Discovery Suite and its use as retrieval, exploration and collaboration platform in the norm and patent domain. Semantic data harmonization, knowledge extraction and disambiguation, semantic search and graphical analysis are key features in order to support the patent officer as well as the research staff. A newly built module for semantic image analysis and exploration, linking semantically enriched metadata and figures to – before – unstructured text, is introduced.

Keywords: semantic search, patent retrieval and analytics, visualization, information extraction, semantic disambiguation, figure search, image analysis.

1 Challenges in Patent and Norm Retrieval

Patents originate from different sources and consequently are heterogeneous in structure, content and format [2]. Manual analysis requires substantial effort (time and resources), special knowledge about the domain and coordination work. Common tasks in the patent retrieval domain include concept based search without having knowledge of the exact textual representation in the patent text, searching for information that supports an appeal against a given claim and search for hidden matters, semantic meanings and relations in large data sets. Also, the analysis of topics and trends within a certain technological field over time and the continuous surveillance of literature and patent publications in the context of competitive intelligence tasks are crucial. As retrieval and analysis tasks can be huge and involve different stakeholders potentially at different places or even countries and with differing domain know-how, support for collaborative search is sought.

Search tasks are also time consuming as crucial information is frequently hidden in drawings by patent applicants (an obfuscation technique). Due to the lack of

M. Salampasis and B. Larsen (Eds.): IRFC 2012, LNCS 7356, pp. 89–93, 2012.
© Springer-Verlag Berlin Heidelberg 2012

automatic image exploration tools, searching for information within a large set of patent images remains a mostly manual task. Given the large number of drawings[1], methods for indexing and searching patent images would tremendously help in the patent examination process. In patents, texts and images are linked through references in the patent text itself. This kind of association is not represented explicitly in patent databases, so it is not possible to make efficient use of them during search. Also, information regarding the type (e.g. technical drawings, diagrams, flow charts and graphs) of the images is not available. In many patent examination tasks, however, it is important to focus on a specific image type. For example, images showing graphs are analyzed with respect to the labeling of the axes, the legend, etc. In flowcharts, for example, the components relevant for examination are actions, decision points and connections.

Conventional search tools are often insufficient as they lack support for searching concepts instead of character strings and generally lack explorative means to target large data sets and discover hidden knowledge. Also, extraction of metadata (for example images, document classes, claims, technical devices, measurements) and visualizations with corresponding unstructured data as well as support for collaborative work are mostly missing. We present the m2n Knowledge Discovery Suite, which addresses these issues in a single flexible solution by providing support for efficiently performing patent and norm search, exploration and analysis tasks. As part of the industry track, the focus of this paper is to describe the building blocks and user interfaces of the novel system rather than to scientifically evaluate the performance.

2   m2n Knowledge Discoverer for Norm and Patent Exploration

The m2n Intelligence Management Framework is built on semantic technologies, a feature-rich knowledge discovery toolbox and the concept of model based development. Flexible and ready to use modules are completely configurable based on semantic application models and can be combined to powerful research, analysis and collaboration tools with minimum modeling effort. These modules together are bundled in the m2n Knowledge Discovery product line. Knowledge discovery methods serve to extract information from unstructured data, enhance semantic data models and provide tools for linguistic text analysis, text based search, continuous surveillance of data sources using machine learning methods, filtering facets and topic clouds, document and meta data preview, graphical analysis, workflow support and collaboration additions (sharing of search context, citations, boilerplates, etc.).

The m2n Knowledge Discovery Suite is the basis for enterprise search and retrieval solutions as well as for domain specific knowledge discovery applications like the norm and patent exploration toolset.

Two specific systems in this domain serve as examples, one being the Semantic Norm Annotator and Discoverer for DIN Software [5], which focuses on automated

---

[1] The average number of drawings per patent is 8 in IPC class F – Mechanical engineering, which was included in the system for demoing purposes.

metadata extraction and semantic retrieval functionality. The other system provides the application base of IMPEx[2], a research project focusing on image processing within the patent domain. Key features for both systems are:

- **Semantic data harmonization.** Logic based data models allow semantic integration of heterogeneous (structured and unstructured) data sources into a single semantic data network enhanced by a reasoning engine to infer new information.
- **Knowledge extraction.** Existing machine interpretable metadata is enriched and new metadata created by means of automated document classification, named entity recognition, semantic disambiguation and semantic reasoning. The (extracted) semantic concepts can be used for advanced filtering, data analysis data and the interpretation of interrelations. Thus, it extracts and interlinks information that has not been accessible before by feasible means.
- **Semantic search.** The system analyzes processed source data and automatically deduces term concepts from this data in combination with the semantic data network. This empowers finding information without knowing the exact search term. Thus, the user no longer searches only for character strings but automatically also for abstract concepts (abbreviations, synonyms, related terms), significantly boosting search quality (based on tester and user feedback).
- **Graphical analysis and visual exploration.** Semantic metadata as well as document similarity can be represented graphically for a fast and intuitive approach towards data exploration. Graph visualizations present interrelations, similarity maps and flow visualizations present proximity of search hits in regard to contents and show temporal changes in the data pool. Fig. 1 shows an analysis cockpit for norm exploration, used to gain insight into the structure and content of huge data sets.

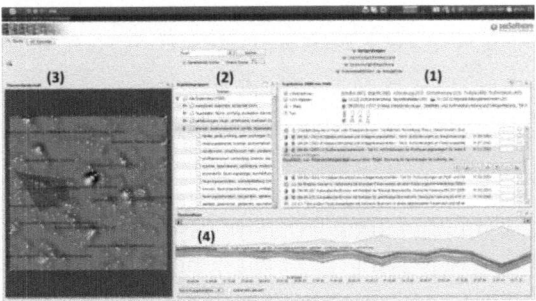

Fig. 1. Analysis of large data sets with the m2n Knowledge Discovery Suite [4]. The screenshot shows the current result list (1), thematic clustering of the data (2), a similarity map (3) and a flow visualization of temporal changes among the thematic clusters (4).

2.1   Image Analysis

The IMPEx project additionally focusses on image processing and linking textual references to images in order to address the before mentioned issues of information

---

[2] FFG FIT-IT project (No. 825846), see http://www.joanneum.at/?id=3922

hiding. More intuitive ways of accessing information in images are provided and images become searchable and semantically comparable by the m2n Knowledge Discovery Suite.

The image processing modules create semantic metadata and provide access to information hidden in images through a combination of optical character recognition (OCR) and various image processing technologies. With the use of OCR, typed and handwritten text and label references are extracted and automatic interlinking between the patent's texts and drawing parts is established.

Advanced image processing technology is applied for segmentation of images into meaningful parts, such as individual figures. For that purpose, morphological operations on pixel level, connected components analysis and subsequent clustering based on the steepest descent of the components' size are applied.

Further, the type of the patent images is automatically identified. A variety of content based features for black and white images and SVM classifiers are applied [3]. Results on a challenging dataset with 9 different image types show a classification accuracy of around 70%. These classification results are also used to validate text-based classification/annotation of the drawing type.

2.2   Accessing Data

The system provides several means of accessing the enhanced and expanded set of semantic enriched metadata in an integrated user interface. One option is the graph visualization module shown in Fig. 2. A given norm's metadata is presented as graph, combining loaded (red, dashed border) and newly extracted (blue) and semantically enriched metadata. It can be used for searching, filtering, data validation and combination from previously plain norm documents.

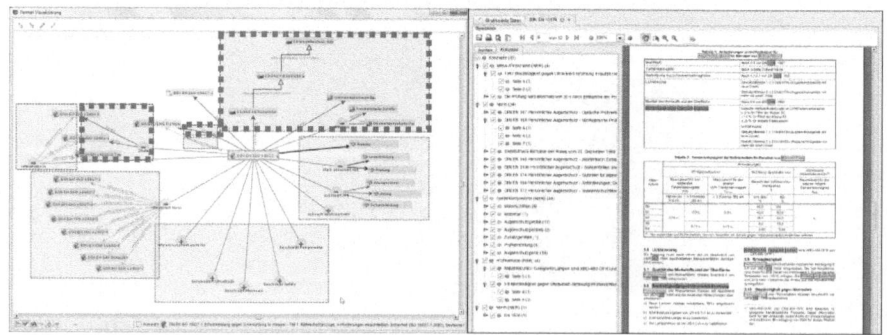

Fig. 2. Exploring existing (red, dashed) and extracted (blue) metadata in DIN Software norm discoverer (graph visualization and document viewer) as presented at [4]. Best viewed in color.

Additionally, the new semantic metadata can be exported to arbitrary databases in predefined formats. Concerning images, a main focus is to make figures findable. Thus, images are segmented into logical figures and indexed by defining findable objects that include figure metadata (for example the image type), text in images and referencing context in patent text. Also, a patent explorer serves to highlight metadata found in (patent) text and to link, for example, the automatically extracted figure references to images and their context in place or to interlink norm references in a specific semantic context (e.g. "used with", "not applicable with"). Fig. 3 shows the viewer with highlighted figure references and linked in place figure preview.

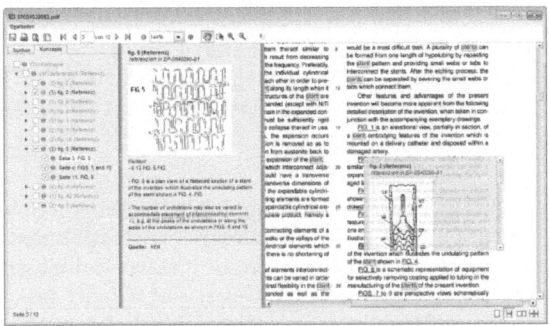

Fig. 3. Patent explorer linking textual references and images

3   Summary

This work shows how semantic technology in combination with text mining and image processing methods is used to assemble powerful retrieval and analysis solutions. Key features include concept based search, semantic data harmonization, semantic metadata extraction, image analysis and integrated user interfaces for exploration. The m2n Knowledge Discovery Suite for norm and patent exploration demonstrates how large datasets are made easily accessible to the user – semantically combining information encoded in text as well as images.

Acknowledgement. The image analysis part of this work was supported by the Austrian Research Promotion Agency (FFG) FIT-IT project IMPEx Image Mining for Patent EXploration (No. 825846). Various Knowledge Discovery Methods have been developed with the Know-Center Graz in the context of the COMET program.

References

1. Information concerning the m2n Knowledge Discovery Suite, http://www.m2n.at
2. Adams, S.R.: Information sources in patents. Walter de Gruyter (2006)
3. Mörzinger, R., Horti, A., Thallinger, G., Bhatti, N., Hanbury, A.: Classifying Patent Images. In: Conference on Multilingual and Multimodal Information Access Evaluation (CLEF 2011), Amsterdam, NL (2011)
4. Schacht, M., Ipsmiller, D.: Informationsextraktion und Ontology Population in der Praxis: Mehrwert für die Inhaltserschließung am Beispiel von Normendokumenten. In: 11th International Conference on Knowledge Management and Knowledge Technologies (I-KNOW 2011), Graz, AT (2011) (technical talk)
5. Schacht, M., Hertel, L., Oberbichler, B., Koch, H.: Nutzung semantischer Werkzeuge. DITRinfo 72, DE (2012), http://www.dinsoftware.de

Hierarchical Classification of Web Documents by Stratified Discriminant Analysis

Juan Carlos Gomez and Marie-Francine Moens

Katholieke Universiteit Leuven
Department of Computer Science, Celestijnenlaan 200A, B-3001 Heverlee, Belgium
{juancarlos.gomez,sien.moens}@cs.kuleuven.be

Abstract. In this work we present and evaluate a methodology to classify web documents into a predefined hierarchy using the textual content of the documents. The general problem of hierarchical classification using taxonomies with thousands of categories is a hard task due to the problem of scarcity of training data. Hierarchical classification is one of the rare situations where, despite the large amount of available data, as more documents become available, more classes are also added to the hierarchy. This leads to a lack of training data for most of the categories, which produces poor individual classification models and tends to bias the classification to dense categories. Here we propose a novel feature extraction technique called Stratified Discriminant Analysis (sDA) that reduces the dimensions of the text-content features of the web documents along the different levels of the hierarchy. The sDA model is intended to reduce the effects of scarcity of data by better grouping and identify the categories with few training examples leading to more robust classification models for those categories. The results of classifying web pages from the Kids&Teens branch of the DMOZ directory show that our model extracts features that are well suited for category grouping of web pages and representation of categories with few training examples.

Keywords: hierarchical classification, text content, web mining, discriminant analysis.

1 Introduction

In the current information era, classification of documents and digital content plays a very important role in many information management and retrieval tasks. On the Web, the number of documents created daily is huge and there is a demand for organizing them in some manner. In this sense, one valuable approach for the management of information is to have the web documents organized by means of hierarchies of conceptual categories, since the use of such hierarchies simplifies the abstraction and facilitates the access to information [8]. In this direction, most of the famous browsing directories (e.g. ODP, Yahoo!) give access to web documents that are manually classified inside a hierarchy of categories. Nevertheless, as the Web changes and grows exponentially, the existing number of non-classified web documents largely exceeds the number of these manually

M. Salampasis and B. Larsen (Eds.): IRFC 2012, LNCS 7356, pp. 94–108, 2012.

labeled inside the hierarchies of the systems. Then it is easy to imagine building automatic classifiers in order to reduce the amount of manual labor on such task. These automatic classifiers, could help to maintain, update and expand web directories [8]; to build customized and relevant views of such directories depending on the necessities of the user, improving the accessibility of content; to improve the quality of content search by returning clustered results [7][9]; and to build efficient crawlers on search engines, among other tasks [25].

In hierarchical classification (HC), categories are arranged as nodes in a graph which represents a hierarchy, usually in the form of a tree. The problem of HC is then defined as: given a hierarchy of categories $\mathfrak{H}$, to automatically classify web documents inside $\mathfrak{H}$ by assigning to the documents a set of categories in a top-down manner, from most general to most specific topics. In general, it is assumed that when a document is categorized with a certain set of node categories from the hierarchy, the document belongs as well to all the node categories along the path connecting the assigned nodes to the root of the graph.

HC has been studied by several researchers in the past years, developing interesting methods ([1][3][5][6][7][8][19][21][29]). In general there are two approaches to solve the problem [26][29]: the top-down approach (indeed the HC approach), which exploits the hierarchy by dividing the problem into smaller ones; and the flat approach, which directly categorize the documents, ignoring the hierarchy. The HC approach simplifies the problem, but it tends to propagate errors made at higher levels of the hierarchy. On the other hand, the flat approach does not propagate errors but tends to be expensive in time for large hierarchies with thousands of categories. In both cases, there is the problem of *scarcity*, where for large web directories the number of documents could be huge but as well is the number of categories [15][19][23]. Then, most of the categories contain a small number of labeled documents, offering little information for training discriminative classifiers. In summary, despite the efforts involved in HC the general performance achieved for this task is still far lower than the one achieved in many other document classification tasks [19].

In this paper we propose a novel technique called *Stratified Discriminant Analysis* (sDA) that deals with the previously described drawback of scarcity of training examples in HC. Our model is focused on *feature extraction* from the text content of web documents. Here, rather than using the direct word distributions over documents and categories, and then train a set of classifiers, we advocate an initial dimensionality reduction of the data, based on Linear Discriminant Analysis (LDA) performed in a novel stratified way.

Flat LDA has been used for a while in text classification tasks [27], but as far as we know, its potential in HC has not yet been explored. In our sDA model, during the training phase, we perform LDA over the set of node categories in each level of the hierarchy, producing a set of projection matrices. Then, when training a node, we use the corresponding projection matrix to transform the training examples depending in which level the node is present. During testing we perform a similar procedure of feature extraction. Before a new web document is classified by a node, it is transformed using the transformation matrix of the corresponding level in the hierarchy. The sDA model is intended to help reduce the dimensionality of the data by keeping the most discriminant features, which could be seen as the "core" information of each category that distinguishes and groups the category from other categories on the same level of the hierarchy. This method could improve category representation, especially for categories with few training examples, which is a common case in most of the categories of a hierarchy.

We test our sDA model on the data and taxonomy underlying the K&T dataset, extracted from the Kids&Teens branch of the DMOZ directory. The Kids&Teens branch embraces a diversity of topics and presents similar distributions on categories and documents as the complete DMOZ directory, which makes the K&T dataset a suitable collection for experimentation. In order to have a better overview of the performance of sDA, we present comparisons with two models: a flat full-term classification model (FFT) and a hierarchical full-term classification model (HFT). As indicated, both models use the whole set of terms and its direct distributions for building the classifiers. The use of direct word distributions for representing documents has proved to have good performance for HC tasks [1][4][19].

The first contribution of our work is the evidence of the feasibility of implementing LDA in a stratified way (sDA) using a decomposition per level, allowing the extraction of reduced meaningful features, which can be used along the hierarchy. The second contribution is the evidence that features extracted by the sDA technique behave very well for representing categories with few training examples in a HC task, and contribute to more robust classification models in those categories. Our findings open new opportunities for feature extraction and HC in general.

The rest of the paper is organized as follows. In section 2 we recall the related work on hierarchical classification. In section 3 we briefly describe the theory behind discriminant analysis. In section 4 we present the complete model for sDA for hierarchical classification of a web document using its textual content. Section 5 illustrates the dataset and setup used in our experiments and discusses the empirical performance of our method in comparison with the baselines. Section 6 is devoted to conclusions and ongoing research.

2   Related Research

HC has been studied in the research community for several years, and there are interesting works devoted to this task. A good survey of the work in this area is presented by Silla and Freitas in [26].

As pointed out before there are two general approaches when studying the HC problem [26][29], namely the flat approach and the HC approach. In the flat approach, one tries to directly categorize documents without making use of the category hierarchy. Despite the fact that a flat classifier could work for HC [4][20], the task of building and testing flat classifiers for large hierarchies tends to be expensive [19].

The HC classification model has been designed using several approaches. In [21], McCallum et al. proposed a hierarchical classification approach using a shrinkage method, in which smoothed parameter estimation of a data-sparse child node is used with its parent node in order to obtain robust parameter estimates. In [6], the authors presented a method based on the Fisher discriminant to rank terms and then perform feature selection by choosing only a given set of the top most discriminant words and applying this method over one part of the Yahoo! hierarchy. In [7] and [8], Dumais and Chen proposed a classifier based on SVM on the top-two levels of the LookSmart directory, called hierarchical SVM (hSVM). In [3], Cai and Hofmann proposed a hSVM based on discriminant functions that are structured in a way that mirrors the category hierarchy, obtained a superior performance than a flat SVM. Another hSVM was used on a very large-scale hierarchy in [19] over the Yahoo! directory, using a selection of local features and direct word distributions, where the hSVM uses a binary SVM classifier per node in the hierarchy, following a Pachinko-model [17][22]. Despite the relative good results obtained with the hSVM, the authors pointed out the difficulties in applying text categorization algorithms to web hierarchies, mostly because of the *data scarcity problem* in the rare categories of taxonomies, where most categories, especially at lower levels of the hierarchy, only have a few examples for training. Xue et al. [29] proposed a two-stage approach for HC. In a first stage, a set of category candidates are estimated for each test document. Then in a second stage, a classification model is trained using only the subset of candidate categories. This approach reduces error propagation by considering fewer categories and improves the performance of a baseline hSVM. Nevertheless, it is necessary to build a new classifier for each test example, and even if this is done in a reduced set of categories the resulting computation cost is very high. Finally Bennett and Nguyen in [1] presented a model based on hSVM but extending it to include a refinement method in order to reduce the effects of error propagation and complex decision surfaces in nodes at the top of the hierarchy. They first integrate the predictive behavior on the training data to change the training distribution in lower nodes; and they integrated a refined expert, which includes a rough class label guess in the lower-level of the hierarchy and propagates this information further up in the hierarchy. The authors report improvements compared to the results of a baseline hSVM using direct word features.

None of the above works exploit the structure of the category hierarchy for the task of feature extraction, which we propose in this paper. Rather than using the whole set of word features with their direct distributions over documents/nodes and then train the model, our approach advocates an initial dimensionality reduction of the data, based on Linear Discriminant Analysis (LDA) performed in a stratified way (sDA), with the purpose of combining all the word features into a new reduced set of attributes. The idea behind this is that sDA would produce a richer set of attributes to represent, identify and group categories than direct word distributions at the different levels of the hierarchy. These extracted core features might help especially in categories where limited training data are available, which is a very common case for categories inside a hierarchy, and where it is difficult to build good classification models.

In flat text classification LDA has been popular over the years as a classification/dimensionality reduction technique [14]. LDA uses the class information to project the training/testing data into a new space where the ratio of between-class-variance to within-class-variance is maximized in order to obtain adequate class separability. Torkkola [27] was one of the first authors to use LDA for text classification purposes. Since then, LDA in text classification has been used following several approaches: using an initial decomposition like Latent Semantic Indexing (LSI) or QR decomposition (QR) and then improved by the LDA model (LSI+LDA and QR+LDA) [14][24], Generalized Discriminant Analysis (GDA) [13][18] and Regularized Discriminant Analysis (RDA) [31]. Cai et al. [2] have proposed a new implementation for LDA, called Spectral Regression Discriminant Analysis, which is able to be linearly scalable for large datasets.

In this paper we focus solely on the effect of extracting discriminative features for the classes in the different levels of the hierarchy using sDA. We expect in this way to better group and identify the individual categories along the different levels of the hierarchy, especially the ones with very few labeled documents, reducing the effects of scarcity. For the classification itself, we rely on a top-down approach similar to a traditional hSVM. In this way our novel sDA model is complementary to other HC approaches described above. In future implementations we can combine the proposed feature extraction method in different classification models like the two-stage approach of [29] or the refined experts of [1], among others, in order to tackle other issues of HC like error propagation and complex decision surfaces at the top levels. To the best of our knowledge feature extraction based on the proposed sDA has not been researched in a HC task.

3   Linear Discriminant Analysis

LDA[1] is usually defined as a transformation of data points, where the transformed data points in the new space present a good separation between classes and good proximity within each class. In this section we give a brief explanation of the main idea behind LDA.

Let $\mathbf{F} = \{(\mathbf{x}_1, \mathbf{y}_1), (\mathbf{x}_2, \mathbf{y}_2), \ldots, (\mathbf{x}_m, \mathbf{y}_m)\}$ be a set of m examples with their corresponding classes, with $\mathbf{X} = \{\mathbf{x}_1, \mathbf{x}_2, \ldots, \mathbf{x}_m\}$, where $\mathbf{x}_i \in \mathbb{R}^p$ is the i-th example, represented by a p dimensional row vector; and $\mathbf{Y} = \{\mathbf{y}_1, \mathbf{y}_2, \ldots, \mathbf{y}_m\} \subset \mathbf{C}$, where $\mathbf{y}_i \subset \{1, \ldots, c\}$ is the set of labels for example $\mathbf{x}_i$ and $c = |\mathbf{C}|$. Then the m documents, with p features, are grouped in c classes.

The goal of LDA is to find a linear transformation $\mathbf{A} \in \mathbb{R}^{p \times l}$ that projects each example vector $\mathbf{x}_i$ to $\mathbf{z}_i = \mathbf{x}_i \mathbf{A}$, where $\mathbf{z}$ is the projected example in a l dimensional space ($l \ll p$), and where the class separability is intended to be maximized. In order to do so, LDA maximizes the following equation

---

[1] Not to be confused with Latent Dirichlet Allocation, a topic distribution model.

$$a^* = \arg\max_{\mathbf{a}} \frac{\mathbf{a}^T \mathbf{B} \mathbf{a}}{\mathbf{a}^T \mathbf{W} \mathbf{a}} \tag{1}$$

where $\mathbf{a}$ is a projection vector inside the matrix $\mathbf{A}$, we then look for a set of l projection vectors. In the previous equation we define the between ($\mathbf{B}$) and within-group ($\mathbf{W}$) covariance matrices as

$$\mathbf{B} = \sum_{k=1}^{c} m_k (\mu^k - \mu)(\mu^k - \mu)^T \tag{2}$$

$$\mathbf{W} = \sum_{k=1}^{c} \left(\sum_{i=1}^{m_k} (\mathbf{x}_i^k - \mu^k)(\mathbf{x}_i^k - \mu^k)^T \right) \tag{3}$$

where μ is the total example mean vector, m_k is the number of examples in the k-th class, μ^k is the average example vector of the k-th, and $\mathbf{x}_i^k$ is the i-th example in the k-th class. Defining $\mathbf{T} = \sum_{i=1}^{m} (\mathbf{x}_i - \mu)(\mathbf{x}_i - \mu)^T$ as the total covariance matrix we have $\mathbf{T} = \mathbf{B} + \mathbf{W}$. Then, equation 1, using a complete set of l projection vectors, can be expressed as

$$\mathbf{A}^* = \arg\max_{\mathbf{A}} \frac{tr(\mathbf{A}^T \mathbf{B} \mathbf{A})}{tr(\mathbf{A}^T \mathbf{T} \mathbf{A})} \tag{4}$$

with $\mathbf{A} = [\mathbf{a}_1, \ldots, \mathbf{a}_l]$ and where tr denotes the matrix trace. The optimization problem in (4) is equivalent to finding the l eigenvectors $\mathbf{a}$ of the following generalized eigen-problem associated with maximum eigenvalues

$$\mathbf{B} \mathbf{a} = \lambda \mathbf{T} \mathbf{a} \tag{5}$$

Since the rank of $\mathbf{B}$ is bounded by $c - 1$, there are at most $c - 1$ eigenvectors corresponding to the non-zero eigenvalues. With the purpose of avoiding the singularity problem of matrix $\mathbf{T}$ (a singular matrix is not invertible), which could occur when the number of features is larger than the number of documents, and to cope with large datasets, in our sDA model we employ the LDA version implemented as the Spectral Regression Discriminant Analysis by Cai et al. [2].

4   Stratified Discriminant Analysis

In order to apply the discriminant analysis for the problem of HC of web documents, we follow a tiered approach and that is why we call our method Stratified Discriminant Analysis (sDA). Since in the actual classification we use a top-down approach for training/testing, each node is trained using examples of its descendants and we could say this step is performed and grouped per level. Hence, with the sDA feature extraction technique we intend to first better group and distinguish the categories at the different levels of the hierarchy by compressing the documents content in less but very meaningful features.

Given a hierarchical structure $\mathfrak{H}$, defined as a tree for this work, and a set $\mathbf{F} = \{(\mathbf{x}_1, \mathbf{y}_1), \dots, (\mathbf{x}_m, \mathbf{y}_m)\}$ of web documents, with $\mathbf{X} = \{\mathbf{x}_1, \mathbf{x}_2, \dots, \mathbf{x}_m)\}$, where $\mathbf{x}_i \in \mathbb{R}^p$ is the i-th document, represented by a p dimensional row vector, and $\mathbf{Y} = \{\mathbf{y}_1, \mathbf{y}_2, \dots, \mathbf{y}_m\} \subset \mathbf{C}$, where $\mathbf{y}_i \subset \{1, \dots, c\}$ is the set of labels for the example $\mathbf{x}_i$, with $c = |\mathbf{C}|$. $\mathbf{C}$ is the set of categories in $\mathfrak{H}$.

The general sDA training process is as follows. Considering that $\mathfrak{H}$ is a tree, it is composed by a number (depth) h of levels. In each level of $\mathfrak{H}$ there exists a number c_i of categories $i = 1, \dots, h$. Given a level i in the hierarchy, we group the descendant categories (if the node is not a leaf) into the superior nodes of that level, in such a way that each node in the hierarchy belonging to that level i is considered together with all the examples that belong to such node or its descendants. Once the examples are grouped per category at level i, we then apply a discriminant analysis to compute the transformation matrix $\mathbf{A}_i$, which has at most $l_i = c_i - 1$ column dimensions. This is illustrated in figure 1a. In this figure, $h = 3$ and there are three transformation matrices computed $\mathbf{A}_1, \mathbf{A}_2, \mathbf{A}_3$, with column dimensions of 3, 7 and 7, respectively.

Once we have the transformation matrices, we then transform the training examples in the level i using the corresponding matrix $\mathbf{A}_i$, obtaining a set $\mathbf{Z}_i$ of transformed examples, with column dimension l_i, grouped in the categories of level i. The last step is then to train a binary classifier in every node of level i. For training, we take the examples belonging to a given node (and its descendants) as the positive class and the examples belonging to its siblings (and their descendants) as the negative class, as specified in the siblings policy in [10]. This is illustrated in figure 1b. The process of grouping, applying discriminant analysis, transform examples and train classifiers, is repeated in each level i of $\mathfrak{H}$, with $i = 1, \dots, h$.

When assigning categories to a new, unseen, document during the testing phase, the sDA model first computes the desired transformation of the example in all the levels of the hierarchy, later a top-down approach using a Pachinko-model [17] [22] is employed: first classifying the document at the uppermost level and then for each binary classifier that "fires" (predicts positive), classifying the document at

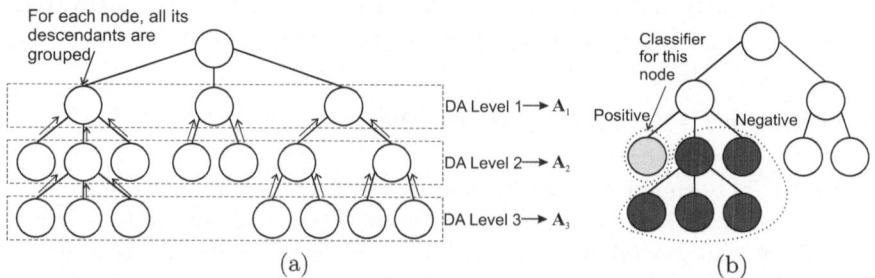

(a)   (b)

Fig. 1. Computation of discriminant analysis per level of the hierarchy (a), grouping the descendant examples into each parent category. Training a classifier for a given node of the hierarchy (b), the light grey nodes indicate the positive category and the dark grey nodes indicate the negative category.

the next lower level, and changing the representation of the document by using a new projection according to the level. Since categories are not completely disjoint, documents can belong to multiple children nodes, having a natural multi-label assignment. In our model we work with mandatory leaf-node predictions, i.e. a complete path from the root to a leaf node must be predicted. In cases where none of the children classifier assigns a category, our model assigns the most likely category according to the probabilities estimated by the classifiers, assuring a full path classification. If there are labeled documents in a non-leaf node, we insert for such non-leaf node a new child node, and all of these labeled documents on the non-leaf node will be included in the new child node. In this way, all documents are placed in leaf-node categories. The leaf mandatory prediction was imposed since in the K&T dataset, most of the nodes are leafs. However, such restriction could be relaxed for other datasets or applications; stopping the classification of the new document in a given level, when the probability of assigning a lower-level category is less than a specified threshold.

Using the notation by Silla and Freitas [26], our sDA method could be defined as the 4-tuple <T, MLP, MLNP, S>, which uses a tree (T) hierarchical structure, allows multiple labels paths (MLP), requires mandatory leaf-node prediction (MLNP), and uses siblings (S) to define the negative examples during training. Further extensions could be easily inserted into the general model.

5   Experimental Analysis

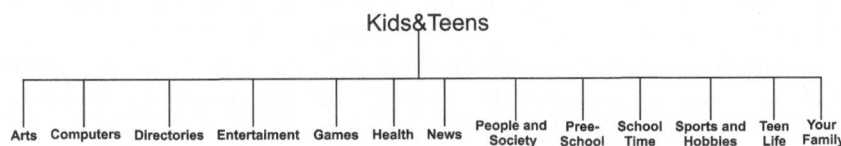

Fig. 2. First level of the hierarchy of the Kids&Teens dataset

5.1   Dataset

In order to test the validity of our sDA model, we perform tests on a dataset (K&T dataset) obtained from the Kids&Teens sub-category of the DMOZ directory[2] from the Open Directory Project (ODP). The hierarchy in the K&T dataset, as in the whole DMOZ directory [29], is a tree, where the longest path (depth of the tree) is 10, considering the Kids&Teens category as level 0, from where all the rest of the categories descend. In order to avoid a large number of non standard character encodings, we decided to eliminate the sub-category Kids&Teens/International. The Kids&Teens hierarchy up to level 1 is shown in

---

[2] http://www.dmoz.org/

Figure 2. It is possible to observe that this hierarchy embraces a large diversity of topics.

In order to extract the content from the web pages we took the title and description of each site from the rdf files provided by the DMOZ directory[3] and then, using the links from the same rdf files, we crawled each site up to level two of the internal links in January 2012. We extracted only the textual content from the web documents, ignoring the HTML tags. As a result, we downloaded a total of 28004 web documents to form the K&T dataset. We considered only the first 40000 characters from each document and we filtered out the stop words. For the sake of statistical significance, we wanted to perform a 5-fold cross validation over the dataset, for that reason we eliminate categories with less than 5 documents. Finally, we filtered out again terms that appear in less than 10 documents, which is a good approach to pre-select informative words [30] and to avoid extra noise. In this way we keep a total of 20639 web documents, containing 43707 unique terms, with 2704 categories in the hierarchy, but with the documents allocated in the 2113 leaf categories. We transformed such documents into vectors using a tf-idf schema and normalized each vector to 1. This normalized representation has shown good results for text categorization using SVM [16].

Several documents in the K&T dataset have multiple labels. Nevertheless, as has been noticed as well for the complete DMOZ directory [23], most of the documents in the K&T dataset are labeled with only one category. There are 20459 documents with 1 label, 169 with 2 labels and only 11 with 3 labels. On average, each document has 1.009 labels. In this case, the almost absence of multi-label documents indicates that no threshold tuning along with classifier training would be required, as suggested by [19].

From figures 3a and 3b it is possible to observe that both categories and documents have peak distributions over levels in the K&T dataset. That is, there are more categories and documents in the middle than at the upper and lower levels of the hierarchy. In figure 3c we see the relation between the number of documents and number of categories. From this graph we observe that over 40% of the categories contain 6 documents or less, and that over 60% of categories contain less than 10 documents. Similar distributions and features have been observed in the complete DMOZ directory [29]. In this sense, the K&T dataset could be considered a good sample of the whole directory.

5.2 Models

As baselines for comparison with our sDA method we use two models constructed with the whole set of word features and its direct distribution (using the tf-idf vectors): a flat full-term classification model (FFT) and a hierarchical full-term classification model (HFT). The FFT model considers all the leaf-nodes as the possible categories, then it trains a binary classifier for each of them, using as positive data the documents belonging the category and as negative data all the remaining training documents, known as the *all* policy in [10]. During testing, it

---

[3] Available at: http://www.dmoz.org/rdf.html

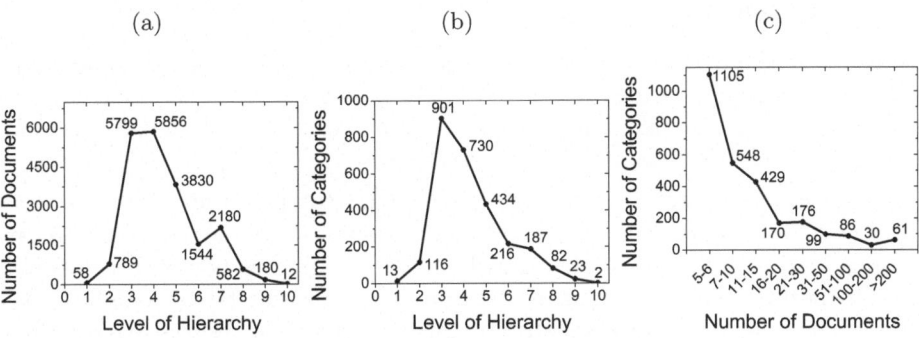

Fig. 3. Category (a) and document (b) distributions in the K&T dataset, and relation of number of documents and number of categories (c)

evaluates each new document for each category and assigns the categories that the classifiers assign as positive. The HFT model is built in a similar way than the sDA, as described in section 4: it trains a binary classifier for every node, taking the examples belonging to a given node (and its descendants) as the positive class and the examples belonging to its siblings (and their descendants) as the negative class. This *siblings* policy has presented good results for HC [10]. During testing a top-down approach is employed: first classifying the document at the uppermost level and then for each classifier that "fires" (predicts positive), classifying the document at the next lower level. The essential difference between the HFT model and the sDA model is in the use of features, i.e. level projected features for the sDA and the whole set of words distributions (tf-idf vectors) for the HFT model.

In the three models we use a linear SVM as the binary classifier in each category node. We use the implementation of the LibLinear [11] classifier, which is especially suited for training SVM in linear time. We use the following settings for the LibLinear: complexity constant C equal to 10000 (for linear SVM on text classification a large C tends to work fine [14]); building of logistic models, in order to obtain probabilities rather than yes/no classifications; and a threshold of 0 (separating hyperplane) for each given binary classifier to determine whether a test document belongs or not to the category. We use the wrapper for LibLinear from the Weka package [12], and the Java LibLinear library[4].

All the models were implemented in Java. As mentioned before, we use the LDA implementation of the Spectral Regression Discriminant Analysis from [2] provided by the authors[5]. We conducted the experiments using a desktop PC with a 3.4Ghz Intel Core i7 processor and with 16Gb in RAM.

---

[4] Available at: http://www.bwaldvogel.de/liblinear-java/
[5] Available at: http://www.zjucadcg.cn/dengcai/SR/index.html

5.3   Performance Measures

To compare the performance of the models, we decided to use the standard label based macro precision, recall and $F1$, which are well understood for text classification in general. These measures are recommended in [28] and [26] for multi-label HC. We use here $Precision = \frac{1}{c}\sum_{i=1}^{c}\frac{TP_i}{(TP_i+FP_i)}$, $Recall = \frac{1}{c}\sum_{i=1}^{c}\frac{TP_i}{(TP_i+FN_i)}$ and $F1 = 2\frac{Precision \cdot Recall}{Precision+Recall}$, computed for each category $i : i = 1,\ldots,c$, and which are then macro averaged. We focus on macro averaging of the performance measures because it estimates how well a model is able to perform along categories, disregarding their size [1] by weighting the performance for each class equally. In this way macro measures avoid to bias the performance towards very dense categories, neglecting or ignoring the scarce ones. Since scarce categories are very frequent in HC (see figure 3c) macro performance is considered as a good statistic to evaluate HC [1][15].

Table 1. Performance of the models in terms of precision (Prec), recall (Rec), F1, training (Tr) and testing (Ts) time. The times are expressed in seconds.

FFT					HFT					sDA				
Prec	Rec	F1	Tr	Ts	Prec	Rec	F1	Tr	Ts	Prec	Rec	F1	Tr	Ts
0.6008	0.6253	0.5916	8150	574	0.6085	0.6262	0.5945	**625**	**46**	**0.6167**	**0.6370**	**0.6035**	1762	189

5.4   Results

The comparative results of the **FFT**, **HFT** and sDA models for the experiments with a 5-fold cross validation on the K&T hierarchy are summarized in table 1. The performance in terms of recall, precision and F1 measures is computed using the *true* most specific categories and all their ancestor categories, and the most specific *predicted* categories and all their ancestors for each web document. The quantities in the table are averages over the 5 folds. We have performed one-sided Wilcoxon signed rank tests for each performance measure across the folds, testing the hypothesis of $F(x) > G(y)$, i.e. if the values of one method tend to be better than the ones of other method. All the differences for the sDA model in comparison with the FFT and HFT models are significant with $p < 0.05$. The differences between the FFT and the HFT models are only significant for precision with $p < 0.05$ but are not significant for recall and F1. That means that the FFT and the HFT models perform similarly for such measures. Nevertheless, the the HFT model performs much better than the FFT model regarding the time complexity. The HFT model performs one order of magnitude faster than the FFT model, both for training and testing. In this sense, the benefit of using a hierarchical version of SVM over a flat version is evident. sDA outperforms both state-of-the-art methods in performance but is slower than the HFT model. The sDA model is more expensive because its training time includes the computation of the transformation matrices and the transformation of examples using such matrices in the different levels of the hierarchy. During testing the sDA method needs to perform the transformation of each testing example along the different levels of the hierarchy using the computed matrices. Nevertheless, we observe that the times are still feasible and less than the ones of the FFT model.

In figure 4 the results of the FFT, HFT and sDA models are split per level of the hierarchy according to the average performance of the 5 folds. The performance per level is computed using the true and predicted categories from the root to the given level. In this sense, it is a cumulative performance and the results of level 10 correspond to the ones shown in table 1, which measures the performance in the whole hierarchy.

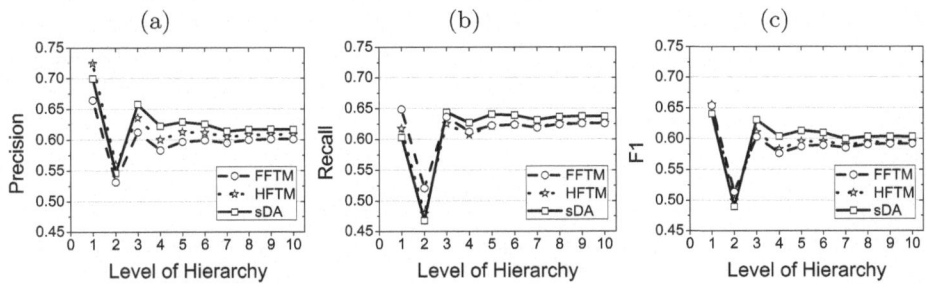

Fig. 4. Performance of the FFT, HFT and sDA models in terms of precision (a), recall (b) and F1 (c) split per level of the hierarchy

The best performance for all the methods is reached at level 1, where there are few categories and most of them contain a large number of examples (given that the examples are collected in the parent nodes). In this level, FFT and HFT models perform slightly better than sDA for F1 (figure 4c), given that with categories at the top level, the effect of a mixture of topics is more evident, and the use of larger vocabularies could better discriminate among them. Similarly at level 2, the FFT and HFT models perform slightly better than sDA because the number of categories is still low and there are more dense categories with overlapping content. At level 2 there is a dip, which could be attributable to the presence of very broad categories like teen_life/online_communities, teen_life/issues, pre_school/people, pre_school/animals, etc. whose content could be assigned to other more specific or descriptive categories.

In level 3 we have the maximum number of categories, and given the general large amount of categories with few training examples (see figure 3c), the sDA model shows its best performance here, meaning that at this level there are many (leaf) categories with few training examples. At levels 4, 5 and 6 we find the majority of the categories. In those levels we still have a large number of scarce categories but less than at level 3. For that reason, the performance of the sDA is better than the FFT and the HFT models but the difference is less in comparison with level 3. As the general number of categories decreases, from level 7, so does the number of scarce categories and the differences in performance between methods does as well, but in general sDA still keeps its better performance than FFT and HFT models.

The effect of scarcity of training examples in categories (i.e. the number of positive labeled documents) on the performance of the methods is shown in figure 5, which plots the precision, recall and F1 depending on the number of training documents in the category. It is clear that for categories with few training examples (between 4 and 6, which represent over 40% of the total number of categories) the sDA model outperforms the FFT model and the HFT model. Especially with only 4 documents, sDA reaches its best performance. This means that the transformation involved in sDA is very well suited to build reduced representations of categories with only a few training examples. As the number of available training documents increases so does the performance of all the models. It is clear as well, that the FFT and HFT models perform better when they use more training documents, since they are able to better generalize with the use of large word distribution features. Nevertheless, on those cases the performance of sDA is still competitive.

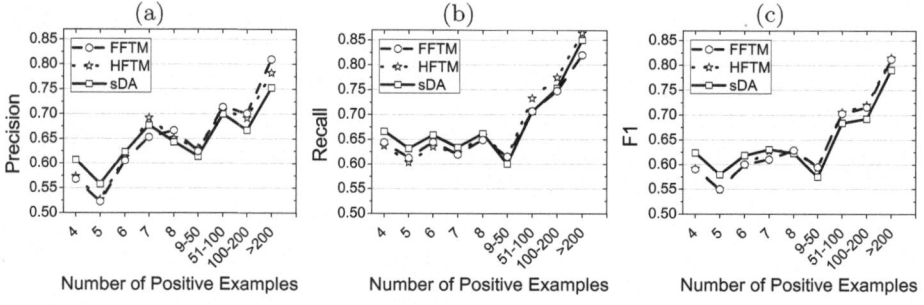

Fig. 5. Performance of the FFT, HFT and sDA models in terms of precision (a), recall (b) and F1 (c) vs. the number of positive examples in the training set

6   Conclusions and Future Work

Our research adds value to a growing body of work exploring how hierarchies structures can be used to improve the efficiency and efficacy of classification of web documents based on their contents. In this article we have presented a method that focuses on a novel feature extraction technique named Stratified Discriminant Analysis (sDA) that exploits the knowledge resided in the hierarchy for extracting compressed features from the word features of the documents.

Our conclusions are that sDA offers advantages over a flat classification model and a hierarchical classification model which use direct word distributions as features. The sDA model reaches a better performance for precision, recall and F1 and keeps the computation time feasible. More specifically, we observed that the sDA model helps separating the different categories in each level of the hierarchy. This model especially improves the grouping and representation of web pages for categories where the number of training examples is scarce. Categories with few training examples are indeed very common in HC and are normally neglected or ignored in benefit of more dense categories. The results presented are truly encouraging and have shown that sDA has the potential to be exploited for the tasks of classifying documents into a hierarchy.

There are many directions to further explore with the sDA model. We are firstly interested into extending the model for larger hierarchies, since the implementation of LDA used in the Spectral Regression model scales well for large datasets, this is feasible. Moreover, with larger hierarchies, the problem of scarcity will become more evident [19], and sDA could be useful. As pointed out, we are interested as well, into combining word features with the features extracted by the sDA model. This could be done by performing sDA over scarce categories and use word features in dense categories; or by linearly combining both type of features into one single model. Another path is the use of sDA in combination with methods that avoid error propagation from top levels, such as the afore mentioned refined experts by [1], the two-stage approach by [29], or the use of different ways of selecting the negative examples during training.

Acknowledgements. This research was supported partially by the CONACYT postdoctoral grant I0010-2010-01/150813, and by the EU FP7-231507 PuppyIR project.

References

1. Bennett, P.N., Nguyen, N.: Refined experts: improving classification in large taxonomies. In: Proc. 32nd ACM SIGIR, pp. 11–18. ACM Press (2009)
2. Cai, D., He, X., Han, J.: Srda: An efficient algorithm for large-scale discriminant analysis. IEEE Transactions on Knowledge and Data Engineering 20(1), 1–12 (2008)
3. Cai, L., Hofmann, T.: Hierarchical document categorization with support vector machines. In: Proc. 13th ACM CKIM, pp. 78–87. ACM Press (2004)
4. Ceci, M., Malerba, D.: Classifying web documents in a hierarchy of categories: a comprehensive study. Journal of Intelligent Information Systems 28(1), 37–78 (2007)
5. Cesa-Bianchi, N., Gentile, C., Zaniboni, L.: Incremental algorithms for hierarchical classification. Journal Machine Learning Reasearch 7, 31–54 (2006)
6. Chakrabarti, S., Dom, B.E., Agrawal, R., Raghavan, P.: Scalable feature selection, classification and signature generation for organizing large text databases into hierarchical topic taxonomies. Journal of Very Large Data Bases 7(3), 163–178 (1998)
7. Chen, H., Dumais, S.: Bringing order to the web: automatically categorizing search results. In: Proc. SIGCHI Conference, pp. 145–152. ACM Press (2000)
8. Dumais, S., Chen, H.: Hierarchical classification of web content. In: Proc. 23rd ACM SIGIR, pp. 256–263. ACM Press (2000)
9. Dumais, S., Cutrell, E., Chen, H.: Optimizing search by showing results in context. In: Proc. SIGCHI Conference, pp. 277–284. ACM Press (2001)
10. Fagni, T., Sebastiani, F.: Selecting negative examples for hierarchical text classification: an experimental comparison. Journal of the American Society for Information Science 61(11), 2256–2265 (2010)

11. Fan, R.E., Chang, K.W., Hsieh, C.J., Wang, X.R., Lin, C.J.: Liblinear: a library for large linear classification. Journal of Machine Learning Research 9, 1871–1874 (2008)
12. Hall, M., Frank, E., Holmes, G., Pfahringer, B., Reutemann, P., Witten, I.H.: The WEKA data mining software: an update. ACM SIGKDD Explorations Newsletter 11(1), 10–18 (2009)
13. Howland, P., Park, H.: Generalizing discriminant analysis using the generalized singular value decomposition. IEEE Transactions on Pattern Analysis and Machine Intelligence 26(8), 995–1006 (2004)
14. Kim, H., Howland, P., Park, H.: Dimension reduction in text classification with support vector machines. Journal of Machine Learning Research 6, 37–53 (2005)
15. Kosmopoulos, A., Gaussier, E., Paliouras, G., Aseervatham, S.: The ECIR 2010 Large Scale Hierarchical Classification Workshop (2010)
16. Lan, M., Tan, C.L., Low, H.-B., Yuan, S.: A comprehensive comparative study on term weighting schemes for text categorization with support vector machines. In: Proc. 14th WWW, pp. 1032–1033 (2005)
17. Li, W., McCallum, A.: Pachinko allocation: Dag-structured mixture models of topic correlations. In: Proc. 23rd ICML, pp. 577–584. ACM Press (2006)
18. Li, T., Zhu, S., Ogihara, M.: Text categorization via generalized discriminant analysis. Information Processing and Management 44(5), 1684–1697 (2008)
19. Liu, T.-Y., Yang, Y., Wan, H., Zeng, H.-J., Chen, Z., Ma, W.-Y.: Support vector machines classification with a very large-scale taxonomy. ACM SIGKDD Explorations Newsletter 7(1), 36–43 (2005)
20. Malik, H.: Improving hierarchical SVMS by hierarchy flattening and lazy classification. In: Proc. Large-Scale Hierarchical Classification Workshop of ECIR (2010)
21. McCallum, A., Rosenfeld, R., Mitchell, T.M., Ng, A.Y.: Improving text classification by shrinkage in a hierarchy of classes. In: Proc. 15th ICML, pp. 359–367. Morgan Kaufmann Publishers Inc. (1998)
22. Mimno, D., Li, W., McCallum, A.: Mixtures of hierarchical topics with Pachinko allocation. In: Proc. 24th ICML, pp. 633–640. ACM Press (2007)
23. Paliouras, G., Gaussier, E., Kosmopoulos, A., Androutsopoulos, I., Artieres, T., Gallinari, P.: Joint ECML/PKDD PASCAL Workshop on Large-Scale Hierarchical Classification (2011)
24. Park, C.H., Lee, M.: On applying linear discriminant analysis for multi-labeled problems. Pattern Recognition Letters 29(7), 878–887 (2008)
25. Qi, X., Davidson, B.D.: Web page classification: features and algorithms. ACM Computing Surveys 41(2), 1–31 (2009)
26. Silla, C., Freitas, A.: A survey of hierarchical classification across different application domains. Data Mining and Knowledge Discovery 22(1), 31–72 (2011)
27. Torkkola, K.: Linear discriminant analysis in document classification. In: Proc. IEEE ICDM Workshop on Text Mining. IEEE (2001)
28. Tsoumakas, G., Katakis, I., Vlahavas, I.: Mining multi-label data. In: Data Mining and Knowledge Discovery Handbook, ch. 34, pp. 667–685 (2010)
29. Xue, G.R., Xing, D., Yang, Q., Yu, Y.: Deep classification in large-scale text hierarchies. In: Proc. 31st ACM SIGIR, pp. 619–626. ACM Press (2008)
30. Yang, Y., Pedersen, J.O.: A comparative study on feature selection in text categorization. In: Proc. 14th ICML, pp. 412–420. Morgan Kaufmann Publishers Inc. (1997)
31. Yen, J., Wang, T.: Regularized discriminant analysis for high dimensional, low sample size data. In: Proc. 12th ACM SIGKDD, pp. 454–463. ACM Press (2006)

PatMedia: Augmenting Patent Search with Content-Based Image Retrieval

Stefanos Vrochidis[1], Anastasia Moumtzidou[1], Gerard Ypma[2],
and Ioannis Kompatsiaris[1]

[1] CERTH-ITI, Thessaloniki, Greece
{stefanos,moumtzid,ikom}@iti.gr
[2] ASML Netherlands BV, Veldhoven, The Netherlands
gerard.ypma@asml.com

Abstract. Recently, the intellectual property and information retrieval communities have shown increasing interest in image retrieval, which could augment the current practices of patent search. In this context, this article presents PatMedia search engine, which is capable of retrieving patent images in content-based manner. PatMedia is evaluated both by presenting results considering information retrieval metrics, as well as realistic patent search scenarios.

Keywords: patents, image search, content-based, retrieval, PatMedia.

1 Introduction

Nowadays, the growing number of patent applications submitted worldwide calls for the need of advanced patent search technologies. Despite the fact that patents contain multimedia information (text, tables, images), most of the current patent search systems still rely only upon text. Recently, both the Intellectual Property (IP) and the Information Retrieval (IR) communities have shown great interest in patent image search expressed with common research activities and with talks in relevant symposiums and conferences (e.g. IRFS[1]). Non-textual elements can play a crucial role in patent search, since image examination is considered very important to patent searchers. It is evident that the tasks of browsing patent images could be speeded up with the aid of search engines, which could retrieve images based on their visual content.

The first attempts towards intellectual property image search were in the field of trademark search. In the field of patent search, one of the first systems was PATSEEK [1], which is an image-based retrieval system for the US patent database, while another system comes from LTU Technologies[2], which was used by the French patent office (INPI). Recently [2] a more holistic patent image processing and retrieval framework was presented.

This paper presents the PatMedia search engine, which retrieves patent images in a content-based manner and comprises the front end of the aforementioned framework.

---

[1] Information Retrieval Facility Symposium (IRFS).
[2] LTU Technologies. http://www.ltutech.com/en/

M. Salampasis and B. Larsen (Eds.): IRFC 2012, LNCS 7356, pp. 109–112, 2012.
© Springer-Verlag Berlin Heidelberg 2012

2   PatMedia Search Engine

PatMedia is a hybrid retrieval engine for patent images, which is capable of retrieving content as in three modes: a) text-based b) visual similarity-based and c) concept-based. The interface of PatMedia[3] (Figure 1) is web-based and includes the left column, where the queries can be submitted, while the results are presented along with additional retrieval options in the main container window.

Textual search is based on the textual indexing of figure captions and references identified in the document, which are automatically linked with the figures in the drawing section by exploiting the figure label information using OCR techniques [2].

Image Retrieval by visual similarity is realized by comparing visual descriptors extracted from patent images. The main characteristic of patent figures is that they are mostly black and white and they depict technical information in diagrammatic form. To cope with the special nature of patent images we employ as features the Adaptive Hierarchical Density Histograms (AHDH) [3], which are specifically designed to deal with such complex drawings.

Concept-based retrieval relies upon a supervised machine learning-based framework, which exploits the aforementioned visual features. During the training phase, we manually select important concepts (e.g. ski boots) inspired by the IPC classification schema and we assign them to patent images, creating in this way a training set. Then, we train a Support Vector Machine (SVM) classifier for each concept using the train data. Finally, we feed every new patent image to each classifier, which assigns a confidence score for the corresponding concept.

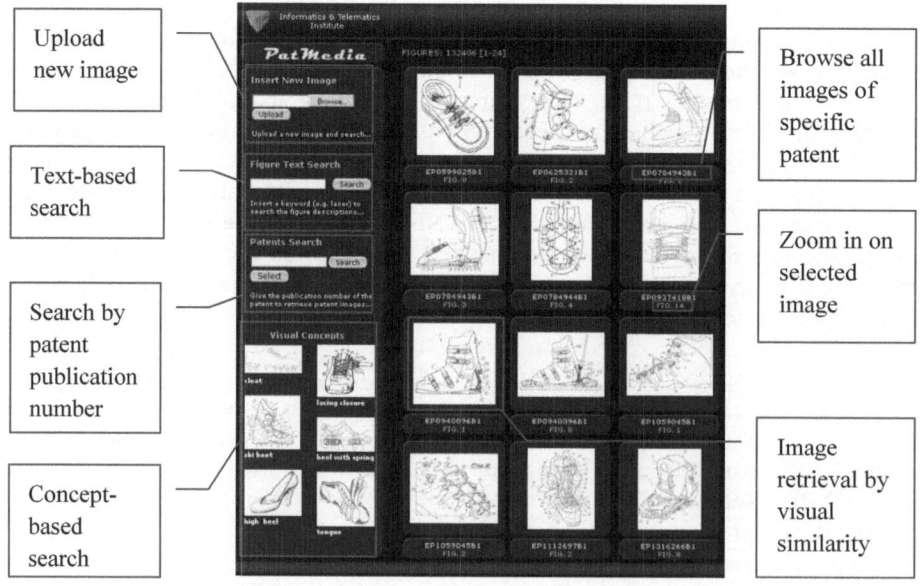

Fig. 1. PatMedia Interface

---

3   Results and Evaluation

3.1   Results and Quantitative Evaluation

In order to evaluate the visual similarity we involved a dataset including 2000 patent images extracted from about 200 EPO patents IPC G11/B7 (optical recording). We have chosen 110 images of diverse content from this collection. For each query image, a relevant response set was identified, containing from 3 to 25 images and we employed a distance threshold, which discriminates relevant from irrelevant images. The results report 77.4% recall for 49% precision. Concerning scalability, PatMedia was tested for over 300000 images and the retrieval time was below 15 seconds.

To evaluate concept retrieval we used a dataset[4] of around 300 patents from A43B and A63C IPC subclasses (parts of footwear) and we extracted around 1000 patent images. The selected concepts and the results are presented in table 1, while table 2 illustrates an example of the Roller skate visual concept.

Table 1. Results for visual concepts

Concepts	Prec.	Recall	F-score
Cleat	84.38%	45.76%	59.34%
Ski boot	84.62%	67.35%	75.00%
High heel	82.69%	72.88%	77.48%
Lacing closure	79.17%	41.30%	54.29%
Heel with spring	69.70%	54.76%	61.33%
Tongue	75.68%	57.14%	65.12%
Roller skates	82.50%	49.25%	61.68%
Average	79.82%	55.49%	64.9%

Table 2. Visual concept example

Roller skate	
A shoe or boot with wheels attached to its sole for skating on hard surfaces IPC groups: A43B5 and A63C17	

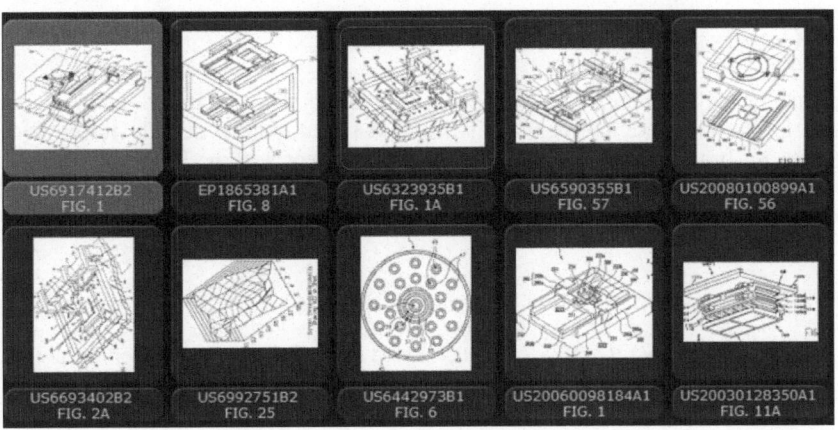

Fig. 2. Image similarity results for Figure 1 in US 6,917,412B2 patent

---

[4] Retrieved from MAREC patent collection, http://www.ir-facility.org/proto types/marec

3.2   Patent Search Scenarios and Qualitative Evaluation

With a view to providing a qualitative evaluation of PatMedia, we have uploaded around 320.000 images from about 15.000 patents from IPC G03F007/20 extracted from MAREC database and performed specific patent search cases. In the first scenario we have searched for similar images with an image included in the US 6,917,412B2. The results (figure 2) include relevant prior art (marked with red) that would have been cited on the search report. In the second use case we have executed an image similarity search with an image that is more commonly used. The results in figure 3 indeed show many very similar images. By ordering these images on publication date we could analyse how this image evolves over time and answer questions such as: when a specific feature is introduced for the first time. These and other similar tasks and results during the qualitative evaluation show that PatMedia is very useful in patent invalidation, patent valuation and competitive intelligence research.

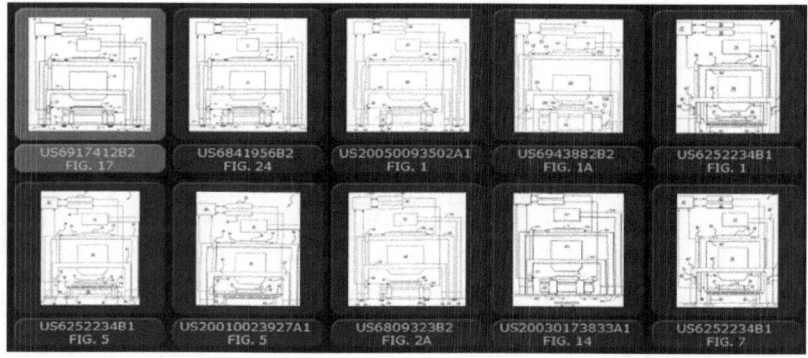

Fig. 3. Image similarity results for Figure 17 in US 6,917,412B2 patent

4   Conclusions

In this paper we have presented PatMedia search engine, which can be used either as standalone tool or complement existing text-based patent search frameworks as powerful, no-brainer check at the start of every patent search. The evaluation shows that the retrieval functionalities of PatMedia can help patent searchers to speed up their tasks and increase the recall of their searches in several cases.

References

1. Tiwari, A., Bansal, V.: PATSEEK: Content Based Image Retrieval System for Patent Database. In: Proc. International Conference on Electronic Business 2004, Beijing, China (2004)
2. Vrochidis, S., Papadopoulos, S., Moumtzidou, A., Sidiropoulos, P., Pianta, E., Kompatsiaris, I.: Towards Content-based Patent Image Retrieval; A Framework Perspective. World Patent Information Journal 32(2), 94–106 (2010)
3. Sidiropoulos, P., Vrochidis, S., Kompatsiaris, I.: Content-Based Retrieval of Complex Binary Images. In: Proc. of CBMI 2010, Grenoble, France, June 23-25, pp. 182–187 (2010)

Generating Variant Keyword Forms
for a Morphologically Complex Language Leads
to Successful Information Retrieval with Finnish

Kimmo Kettunen and Paavo Arvola

School of Information Sciences, University of Tampere, Finland
{kimmo.kettunen,paavo.arvola}@uta.fi

Abstract. This paper discusses information retrieval of Finnish and keyword variation management by generating inflected variant keyword forms. Finnish is a highly inflectional language, and thus keyword variation management of queries and query indexes is of utter importance for successful Finnish full-text retrieval. In the paper we show that generation of a quite small number of variant keyword forms leads to good retrieval performance using a probabilistic best-match retrieval system (Lemur). Generation of almost the full paradigm of inflected nominal forms improves the results slightly. We have also interesting results with regards to different index types: our evaluation shows that generated inflected queries behave extremely well in a lemmatized index, which is supposedly not suitable for this query type. We also show that in a research environment even inexact generation that produces lots of incorrect inflected forms achieves high precision-recall performance without considerable loss in query throughput effectiveness. We use two different word form generators and their variants and compare the results to commonly used reductive word form variation management methods, stemming and lemmatization. The paper includes also a short discussion about usage of the variant keyword method with Web search engines.

Keywords: information retrieval, management of keyword variation, Finnish, Frequent Case Generation.

1   Introduction

One of the basic problems of full-text retrieval is variation of word forms that is caused by morphology of natural languages. Shortly put, this means that one base or dictionary form of a word in language may occur in different (inflected) variant forms in texts. Out of this follows that many times the principle "one keyword–one concept - one match" does hold not in the textual index of retrieval systems due to morphology alone. Therefore something needs to be done to morphological variation so that performance of information retrieval (IR) systems will not suffer too much if the language has a rich or at least medium rich morphology. To overcome the problem of keyword variation several management methods have been proposed during the history of textual IR. One of the first solutions to the problem of

M. Salampasis and B. Larsen (Eds.): IRFC 2012, LNCS 7356, pp. 113–126, 2012.

morphological variation was a procedure called stemming, and the first stemmer for English was introduced in 1968 by Janet Lovins [1]. Since then many stemmers for English and other languages have been introduced and an interested reader might take a look at discussion found for example in Harman [2], Frakes [3], Hull [4], Hollink et al. [5] and Galvez et al. [6]. Stemming has been the most widely applied reductive morphological technique in IR. In stemming distinct variants of word forms are conflated or reduced (optimally) to one form that may be a base form or just a technical stem. An example would be optimal reduction of the set {cat, cats, cat's, cats'} to one lemma {cat}. A more sophisticated reductive keyword variation management method is lemmatization, which produces base or dictionary forms of inflected word forms with rules and a dictionary (e.g. Koskenniemi [7]) or without using a dictionary as in Loponen and Järvelin [8].

Another option for keyword variation management is generation of variant word forms for query keys. In this case variant forms of the keyword are generated using a base form. Generation works simply. Given a base form, like {cat}, the generator produces all the variant forms of it, in this case {cat, cats, cat's, cats'} or just a subset of the forms if the number of variant forms is very high. These produced variant forms are given to the retrieval engine which matches them in the plain inflected text index that has not been analyzed with language technology tools, i.e. the index contains the textual words as such.

In this paper we study the effectiveness of generation with regards to a morphologically complex language, Finnish. We try out two different generators and their versions with different number of generated variant keyword forms. We wish to show that generation of variant keyword forms is competitive with reductive keyword variation management methods, stemming and lemmatization, and also quite error tolerant with regards to false generations. We study also the effect of different base form creation methods by using two different lemmatizers, a dictionary based and a statistical one. We also briefly evaluate the effect of non-standard index type for generated queries. Furthermore we perform a small web search test using Bing and Yahoo search engines with variant keyword forms. Altogether our evaluation is the most comprehensive evaluation of the variant keyword method for Finnish so far.

2   Inflected Keyword Generation for Finnish

Inflectional morphology of Finnish is quite complicated as nouns and adjectives that mostly affect IR results may in principle occur in thousands of forms. Nouns have about 2000 grammatical forms and adjectives about 6000, when also clitics and possessive endings are counted (Karlsson [9]). In practice only a fraction of the forms really exist in running texts (cf. Kettunen, Airio and Järvelin [10]). The optimal query performance boosting number of keyword variant forms is between 6 and 12 forms when using a mechanical case form generation. It could be even less, if the semantics of the keywords was known and only the case forms for the semantic type of the noun could be generated (Kettunen and Airio [11]; Kettunen et al. [10]).

In our tests we rely partly on earlier research concerning generation of Finnish keyword variants. In order to avoid use of too many keyword variants Kettunen and

Airio [11] and Kettunen and colleagues [10] developed a linguistic frequency based method called Frequent Case Generation (FCG). For languages with large or moderately large number of variant word forms, the number of generated forms is restricted to only the most frequent forms that are first determined with the use of corpora. By restricting the number of generated keyword forms to only the most frequent forms Kettunen, Airio and Järvelin [10] have been able to achieve IR results that are 86–95 % of the best available comparison, use of a lemmatizer or stemmer for Finnish, Swedish, German and Russian in the Inquery query system with simulated queries. All of the languages are at least morphologically moderately complex. Automatic generation of FCG style query keywords was proven feasible with English, Finnish, German and Swedish in the Lemur query system in Kettunen [12]. Leturia et al. [13,14] have used the same type of approach, usage of the most common inflected word forms, in their web-search enhancement for the Basque language. Recently, Bengali, Gujarati and Marathi IR has been tried with the method successfully by Paik and colleagues [15]. In this study we adopt both the FCG strategy and usage of a more comprehensive list of variant forms in our evaluation.

3   Generators

We used two different generators and their variants in our evaluation. The generators, SWERG and FCG_12, were implemented by the authors. These generators have not been used in IR evaluation earlier, and we describe them briefly here.

Our two first generators are simple word ending based rule generators SWERG and SWERG+. The SWERG generator is based on rules, derived from a database of 261 087 lemmas and their inflected forms in six cases. More precisely, the generator is taught so that the singular nominative is compared with its plural form and their genitive, partitive, inessive, elative and illative cases. For the extended version of the generator, SWERG+, translative, adessive, ablative, allative, essive and abessive are generated using the inessive forms and replacing the ending of elative "ssa" with the respective endings. The rules are applied to 3052 different lemma endings and the total numbers of individual rules of the SWERG and SWERG+ are 21 395 and 47 966 respectively.

In the generation process, the rules are applied to a lemma in a greedy fashion, so that a lemma may have several form candidates based on endings of varying lengths. For the lemma *horisontti*, for example, the rules are searched for the endings "ti", "tti", "ntti", "ontti", "sontti" and so on. This greedy method delivers a comprehensive set of forms, but also a set of incorrect ones. SWERG+ generates 31 incorrect forms for the lemma *horisontti*, although the correct word within an incorrect form is intuitively distinguishable. It is worth noting that the ungrammatical forms may match to the relevant documents. For example, due to human misspelling or wrong inflecting, a disjunctive query of the generated incorrect forms of the lemma *horisontti* has about 92 matches in Google. Consequently, the benefit, or disadvantage, of generating incorrect forms becomes evident when looking at the effectiveness results.

Our second set of generators, FCG_12 (and its more restricted version FCG_6), are rule-based generators with handcrafted rules. They have been implemented on the basis of our earlier inflectional stem generator in Kettunen and colleagues [16]. We have limited the generation of FCG_12 to six case forms of Finnish nominals - nominative, genitive, partitive, inessive, elative and illative - which are the most frequent case forms of nominals in running Finnish texts and cover 84–88 % of the nominal forms in Finnish texts [11]. FCG_6 produces only the three most frequent case forms, nominative, genitive and partitive, in singular and plural

If we put together the whole generation approach used in this study, we have a following picture: our four generators produce 6–24 variant forms out of the keywords. Six forms is about 21 % of the paradigmatic noun forms (2*14), 12 about 43 % and 24 about 86 %. Thus the generators cover the most frequent range of the variant forms, but also usage of almost a full paradigm – minus two cases, very rare instructive and comitative – will be evaluated.

4   Evaluation

We evaluated our test scenario using one of the standard Finnish collections, CLEF 2003, which was the only one available for us. Finnish CLEF 2003 collection has 55 344 documents and 45 topics with relevant documents, the total number of topics being 60. Our query engine was Lemur (version 3.12). For comparison purposes we used plain word queries, stemmed queries (Snowball Finnish stemmer [17]) and lemmatized word queries (FINTWOL [18]) and respective indexes. Snowball stemmer is a standard stemmer for Finnish, and there are not available many other stemmer options, one exception being Regstems by Kettunen and Baskaya [19]. Regstems has been evaluated as a slightly lighter stemmer performing quite equally with Snowball. FINTWOL is a commercial lemmatizer, which usually performs best in IR evaluations of Finnish.

Our evaluation runs were performed for long (title and description, TD) and short queries (title only, T) with the CLEF 2003 data. We used four different effectiveness measures in the evaluation: mean average precision (MAP), number of returned relevant documents (Rel-ret), precision at fixed rank point, P(10), and R-precision (R-prec) (cf. Sanderson [20]). R-precision is calculated as P(R), where R is the total number of relevant documents in the collection for a particular topic. The fixed rank position measure, P(10), is a user-oriented measure that evaluates the case where user is only interested in a few relevant results on the top of the first result page

Queries for the generated variant keyword form runs were produced in the following manner: we created base form versions out of the CLEF 2003 topics with FINTWOL lemmatizer, added query structure and generated variant forms for every keyword in the query regardless of their word class. This means that there are false generations among the keywords, as the generators expect to have as input only nominal word forms. Also word forms that are left intact by the lemmatizer, are usually given wrong generations if they are in some other form than nominative singular in the query. Our queries were structured following the usual pattern, where variants of the same key are together under Lemur's *#syn* operator.

4.1   Basic Evaluation

We used FINTWOL in two different ways as a lemmatizer. All the lemmas queries included all the analyses given by FINTWOL, first lemma queries included only the first analysis given by FINTWOL, if there were several analyses. Our generator induced queries were also formed using these two different outputs from FINTWOL, and the notation follows the same pattern in the tables. Results of the basic runs for long and short queries are shown in Tables 1 and 2.

Table 1. Results of runs with 1000 retrieved documents with TD queries. Statistical significance test was performed using the Friedman test on MAP results: very significant differences (p < 0.001) vs. other systems are shown beside the MAP in parentheses. Three best systems by MAP are in bold.

TD queries	MAP	Rel-Ret.	P(10)	R-Prec
1. FINTWOL, all	0.4021 (12)	322	0.2889	0.3680
2. FINTWOL, first	0.5145 (3,8,9,10,11,12)	476	0.3444	0.4595
3. Snowball	0.4218 (12)	450	0.2978	0.3904
4. SWERG, all	0.4743 (3,10,11,12)	451	0.3111	0.4351
5. SWERG, first	0.4706 (11,12)	449	0.3111	0.4388
6. SWERG+, all	0.4886 (3,10,11,12)	453	0.3067	0.4684
7. SWERG+, first	0.4775 (11,12)	452	0.3067	0.4540
8. FCG_12, all	0.4487 (12)	435	0.2978	0.4228
9. FCG_12, first	0.4328 (12)	440	0.2978	0.4011
10. FCG_6, all	0.4138 (12)	426	0.2844	0.3874
11. FCG_6, first	0.4101 (12)	433	0.2756	0.3808
12. Plain	0.3158	405	0.2356	0.2889

The results show that lemmatization is the best method for keyword variation management if only one lemma for ambiguous query words is used in lemmatized long queries. Difference of the best generation method to the lemmatized results is about 2–3 per cent units with long queries and about 2 per cent units with short queries. With smaller number of keyword variants the difference increases to 7–10 per cent in long queries and to 3–10 per cent in short queries.

Overall different generations perform well. If inflected queries are generated from all the lemmas given by the lemmatizer, the performance is slightly better than using only one base form. The difference to plain word baseline is material (> 10 %, Sanderson [20]) in long queries with all the generators except FCG_6. In short queries the difference of all the generated query results except one (#11 in Table 2) is material when compared to plain word queries. Six of the generations outperform Snowball stemming with long queries, and the difference varies from 1 to 5 per cent units. With short queries all generated queries outperform Snowball with about 3–7 per cent units.

Table 2. Results of runs with 1000 retrieved documents with T queries. Statistical significance test was performed using the Friedman test on MAP results: very significant differences (p < 0.001) vs. other systems are shown beside the MAP in parentheses. Three best systems by MAP are in bold.

T queries	MAP	Rel-Ret.	P(10)	R-Prec
1. **FINTWOL, all**	0.4525 (3,10,11,12)	442	0.2956	0.4020
2. **FINTWOL, first**	0.4500 (3,10,11,12)	443	0.2933	0.4020
3. Snowball	0.3251	419	0.2467	0.2841
4. SWERG, all	0.4214 (11,12)	438	0.2756	0.3867
5. SWERG, first	0.4125 (12)	435	0.2689	0.3722
6. **SWERG+, all**	0.4301 (10,11,12)	439	0.2822	0.3946
7. SWERG+, first	0.4227 (11,12)	436	0.2756	0.3802
8. FCG_12, all	0.3815 (11,12)	419	0.2600	0.3405
9. FCG_12, first	0.3777 (12)	419	0.2578	0.3405
10. FCG_6, all	0.3625	413	0.2600	0.3377
11. FCG_6, first	0.3561	412	0.2533	0.3288
12. Plain	0.2620	363	0.2200	0.2427

When results of P10 measure are examined, we notice that the differences between generations and lemmatization are smaller. In long queries the difference is 3.3–6 per cent units (mean of all 0.296), and in short queries the difference is 1.3–5 per cent units (mean of all 0.265). Plain word queries are 7.5 and 10.9 per cent units worse than the best lemmatized result.

4.2   Evaluation with Other Base Form Creation Method

Our main tool for query lemmatization was FINTWOL, a standard morphological base form analyzer with a large lexicon. We had also access to a statistical lemmatizer, Stale [8]. We performed query runs where topics were first analyzed with Stale and generated the queries from the base form output of Stale. As Stale is a statistical lemmatizer with no lexicons at use, it usually returns several base form suggestions for input words. Many of these suggested base forms are non-existent forms which will generate non-existent inflected forms. On the other side, Stale may also give valid base forms for words that are not analyzed by a lemmatizer using a lexicon. Personal names or company names, for example, are many times missing from the lexicon of a lemmatizer, but the base forms of the names may be rightly suggested by the statistical lemmatizer. Results of our Stale base form runs – generations from all the suggested lemmas – are given in Table 3.

Table 3. Results of runs where base forms were generated with Stale, mean average precision (MAP)

	TD queries	Change vs. TWOL base form, abs. % units	T queries	Change vs. TWOL base form, abs. % units
SWERG	0.4268	- 0.048	0.4265	+ 0.005
SWERG+	0.4357	- 0.053	0.4258	- 0.004
FCG_12	0.3965	- 0.052	0.3665	- 0.011
FCG_6	0.3908	- 0.023	0.3554	- 0.007

Results in Table 3 show that long queries generated from a base form given by Stale perform 2-5 per cent units below results of those runs when FINTWOL was the base form producer for generation. Short queries perform at the same level with both base form generation sources. The results are fairly good, and show that the statistical lemmatizer, Stale, is able to produce lemmas well. Supposedly the bigger number of proposed and many times erroneous base forms hurts retrieval in the case of SWERG and FCG_12.

4.3   Generated Runs in the Lemmatized Index

One of the ground truths in word form management of IR is that query runs should be performed in an appropriate index. This means that if you have lemmatized queries the runs should be performed in a lemmatized index; stemmed queries should be performed in a stemmed index etc. We did some experiments where we ran generated queries in a lemmatized index instead of plain word index. The lemmatized index should contain only lemmatized words forms plus plain word forms that the lemmatizer was not able to lemmatize either because they are misspellings or lacking from the dictionary of the lemmatizer or due to error in coding of the lemmatizer. These runs gave surprisingly good results, outperforming the results of generated runs in the plain index. The results are shown in Table 4 for different generations.

Table 4. Results of generated runs from FINTWOL's output in the lemmatized index, mean average precision (MAP). Three best systems by MAP are in bold.

	TD queries	Gain compared to plain index search, abs. % units	T queries	Gain compared to plain index search, abs. % units
SWERG, all	0.5223	0.048	0.4474	0.026
SWERG, first	0.5108	0.040	0.4488	0.036
SWERG+, all	0.5221	0.036	0.4471	0.017
SWERG+, first	0.5107	0.033	0.4485	0.026
FCG_12, all	0.5061	0.057	0.4500	0.069
FCG_12, first	0.5113	0.079	0.4479	0.070
FCG_6, all	0.5061	0.092	0.4499	0.087
FCG_6, first	0.5113	0.101	0.4462	0.090

When the results are compared to the results of Tables 1 and 2, we see that generated long queries perform about 3–10 absolute per cent units better in the lemmatized index than in the plain word index. Short queries perform about 2–9 absolute per cent units better in the lemmatized index than in the plain word index. Generated queries achieve also the best MAPs overall in our whole evaluation outperforming even lemmatized queries slightly.

The reason for this good performance is not self-evident. From the quite equal performance of different generators it seems, that only a few case forms are the main cause of this performance boost: best result of FCG_6 is here only about 1 per cent unit behind or slightly better than results of SWERG+, when their differences in Table 1 are clear. A query-by-query result analysis reveals that FCG_6 gains in more individual queries than SWERG. In long queries FCG_6 gets better results with 28 queries in the TWOL index, as SWERG gets only 19. In short queries FCG_6 gets also better results with 28 queries in the TWOL index, as SWERG gets 22. So there is a clear trend of FCG_6 getting the biggest boost on individual query level in the TWOL index, but the reason for this is not self-evident. This would need further examination with thorough index side matching scrutinization.

Same effect was found with generations that were done from statistical lemmatizer Stale's output with all the suggested lemmas from Stale. Results are shown in Table 5.

Table 5. Results of generated runs from Stale's output in the lemmatized index, mean average precision (MAP)

	TD queries	Gain compared to plain index search, abs. % units	T queries	Gain compared to plain index search , abs. % units
SWERG	0.4832	0.056	0.4792	0.053
SWERG+	0.4829	0.047	0.4785	0.053
FCG_12	0.4693	0.073	0.4364	0.07
FCG_6	0.4741	0.083	0.4397	0.084

With these results we notice a 5–8 absolute per cent unit increase in the MAPs overall with both long and short queries. Results are below those shown in Table 4, where generations from FINTWOL's output were run in the lemmatized index, and the difference is most obviously caused by Stale's erroneous base forms suggestions which lead to false generations that hurt retrieval.

4.4   Searching with Variant Keyword Forms in the Web

Our main aim in this paper was to evaluate the effectiveness of variant keyword search in a laboratory evaluation setting. We did, however, perform a small Web search test with Google's, Yahoo's and Microsoft's Bing engine using three, four, six and twelve variant forms of a set of keywords and compared the results to usage of base form only. The keywords were given to the engines in the form "presidenttiehdokas" OR "presidenttiehdokkaat" OR "presidenttiehdokkaan" OR "presidenttiehdokkaiden" OR "presidenttiehdokasta" OR "presidenttiehdokkaita". We used a set of eight keywords: {*sikainfluenssa, velkakriisi, presidentinvaali, presidenttiehdokas, niinistö, haavisto, väyrynen, soini*}. First four are common nouns, last four surnames of presidential candidates in the Finnish presidential elections in winter 2012.

Google was searched from http://www.google.fi/, Yahoo from http://fi.yahoo.com with option "Suomi" (Finnish). Bing's search engine was used from www.bing.com with option "only Finnish results". Google's engine shows an approximation of the time used in searching, Yahoo and Bing do not show time. All give a number of search hits.

Having done the searches with all the three engines we noticed, that Google's hit counts are unreliable and vary oddly, so we don't show these results. Bing's and Yahoo's hit counts seem more stable and reliable, but perhaps they are also merely approximations, not precise counts.

Tables 6 and 7 show the results of Bing and Yahoo searches with the keywords. Figures show number of documents that the engine has found. Change percentage is calculated against base forms counts.

Table 6. Results of Bing search, search date 8 of February 2012

Keywords	Base form search	Search with 3 forms change, % units	Search with 4 forms change, % units	Search with 6 forms change, % units	Search with 12 forms change, % units
sikainfluenssa *'swine flu'*	29 600	30 000 +1.4	30 000 +1.4	31 300 +5.7	32 200 +8.8.
velkakriisi *'debt crisis'*	30 600	28 400 -7.2	28 400 -7.2	28 800 -5.9	31 100 +1.6
presidentinvaali *'presidential elections'*	33 400	214 000 +540.7	238 000 +612.6	237 000 +609.6	259 000 +675.4
presidenttiehdokas *'presidential candidate'*	46 600	108 000 +131.8	181 000 +288.4	197 000 +322.7	199 000 +327.0
niinistö *surname of a presidential candidate*	254 000	282 000 +11.0	284 000 +11.8	287 000 +13.0	289 000 +13.8
haavisto *surname of a presidential candidate*	272 000	295 000 +8.5	298 000 +9.6	301 000 +10.7	304 000 +11.8
väyrynen *surname of a presidential candidate*	123 000	136 000 +10.6	137 000 +11.4	146 000 +18.7	151 000 +22.8
soini *surname of a presidential candidate*	252 000	276 000 +9.5	276 000 +9.5	276 000 +9.5	285 000 +13.1

Table 7. Results of Yahoo search, search date 8 of February 2012

Keywords	Base form search	Search with 3 forms change, % units	Search with 4 forms change, % units	Search with 6 forms change, % units	Search with 12 forms change, % units
sikainfluenssa	21 400	20 400 -4.7	20 400 -4.7	20 600 -3.7	20 500 -4.2
velkakriisi	31 000	27 900 -10.0	27 700 -10.6	27 700 -10.6	30 700 -1.0
presidentinvaali	30 900	200 000 +547.2	211 000 +582.8	216 000 +599.0	231 000 +647.6
presidenttiehdokas	40 400	93 900 +132.4	167 000 +313.4	188 000 +365.3	194 000 +380.2
niinistö	211 000	233 000 +10.4.	229 000 +8.5	94 100 -55.4	99 500 -52.8
haavisto	195 000	221 000 +13.3.	220 000 +12.8	219 000 +12.3	219 000 +12.3
väyrynen	99 000	109 000 +10.1	111 000 +12.1	112 000 +13.1	114 000 +15.2
soini	182 000	195 000 +7.1	196 000 +7.7	196 000 +7.7	201 000 +10.4

As the data in the Tables 6 and 7 show, the increase in document hits was clear with both Bing and Yahoo. With Bing the increase was from 1.4 % to 675 % at best and also noticeable with small number of keyword variants. In Yahoo hits increased also with most of the keywords (except the first two) about 7–500 % even with the smallest number of keyword variants. *Niinistö* with six and twelve forms performed oddly in Yahoo: the number of hits diminished with more forms, when three and four forms gave a small increase in hits over the baseline. Overall there was a clear tendency of increased document hits with both engines when more forms were added to the query, but there was also variation caused by different keywords.

Increase in the number of document hits is in accordance with the results of Bar-Ilan and Gutman [21], who evaluated search capabilities of major search engines with searches of Russian, French, Hungarian and Hebrew. The results showed that linguistic properties of those languages were not handled very well in the major search engines, and native search engines gave far more documents for searches in these languages.

We do not know how Bing and Yahoo exactly cover morphology of Finnish at present, if they do it at all. Our results imply that there is room for improvement in this respect, as the number of found documents increases so much when inflected forms are given in the search beside the base form.

As we have stated, we are interested in the effectiveness of the variant form searches, and not concerned about the efficiency, i.e. time consumption, of the searches. However, our web search experiences suggest that the increase in search time is not very crucial when variant forms are added to the query. This would need further research, but already on the basis of this very informal timing measurement and user experience it seems that current commercial web search engines can handle quite a big number of variant forms satisfactorily.

5   Discussion and Conclusions

In this paper we have shown that generation of variant key word forms yields in successful retrieval with Finnish in a standard text collection using a state-of-the-art statistical best-match query engine, Lemur. We evaluated a wide range of variant keyword forms, the number of variants being between 6 and 24. Our evaluation is so far the most comprehensive testing of the FCG method for Finnish, which can be considered as a prototypical language for this kind of keyword variation management.

Our evaluation section showed that most of the different versions of keyword generation produced effective searches that compared well with both lemmatization and stemming. Stemming was outperformed with almost every version of generation and generation results were not too much behind lemmatization results in many cases. Performance differences to plain word queries were statistically significant with all the generations in both long and short queries. Generation was also shown to be robust with respect to false generated forms and different sources of base forms.

Despite the good results one might question our approach on the following grounds, which can be considered as possible problems of variant keyword form generation. We discuss the possible problems here briefly.

1) There is a too high number of variant forms to be searched, i.e. searches become computationally inefficient. Counter arguments for this are the following:

i) Kettunen and Airio [11] showed that the time increase with more variant keywords is not very severe. When the number of variant keywords per query term increased from 3 to 12 (four times), time used for queries increased only by 20 %. This was measured in a laboratory evaluation setting, but it gives a rough estimate, anyhow. Our small Web engine test in chapter 3.4 suggests also this. Searching of an index tree is not very time consuming computationally, and the search time does not increase linearly with the number of search keys.

ii) number of keywords in real users' searches is low, usually 1-2 (Jansen et al. [22] Jansen and Spink [23]). This leads to a quite reasonable number of search keys for the whole query, if the number of variants is kept at the lower limit. As we have shown, for a inflectionally rich language like Finnish, a limited number of keywords variants works well in retrieval. This was shown both in laboratory evaluation and in section 3.4 with Web search engines.

2) Bad generation or false forms in general cause problems for generated searches. Counter arguments for this are the following:

i) generation is simpler than lemmatization and false forms are not very common with generators

ii) possible false generated forms which may be due to wrong generation or generating from a wrong word class keyword don't hit anything in the index and thus they do not harm the result set. Thus the generator's output quality is not very crucial; it is important that the generator is able to produce the most common inflected forms but the completeness of the word set is not that crucial.

3) Production of base forms for the generator might not be available.

We have used in our evaluation automatized base form generation as a basis for variant keyword form generation. In a user-oriented test scenario the user would be most probably giving the keywords in their base forms, as a native language speaker is able to give the keywords in their base forms. However, it might also be useful to have a lemmatizer in the front end of the search engine analyzing the inputs of the user and suggesting base forms for keywords that are for some reason problematic for the user. These could include e.g. uncommon words such as foreign place names, product names etc. Also compound splitting might be suggested by the lemmatizer, as parts of the compound could be used for generation in compounding languages, such as Finnish, German and Swedish. Generation seems also to be quite robust with respect to the source of base forms, as shown in section 4.2.

Also other arguments pro generation approach can be listed. When keyword variant generation is used, retrieval index can be kept un-lemmatized or un-stemmed, which leads to a higher architectural simplicity of the retrieval system.

Generation approach may be also otherwise meaningful for small languages. Leturia et al. [13, 14] use same type of generation approach for Basque with good results in a web environment. They emphasize the linguistic and cultural meaning of a native search engine for a small language, especially for the younger generation. Bar-Ilan and Gutman [21], on the other side, have shown, that many times native search engines perform better in non-English searches than Google or MSN, which have not been too keen to improve their search engines' linguistic capabilities for smaller languages. Our initial results with web searches of Finnish corroborate this.

For a small and morphologically at least moderately complex language, variant form generation of keywords offers a simple but effective means to improve web searches. Already in the EU area there are quite many small languages that have a rich morphology and need for word form variation management in text retrieval. Good examples of such languages are for example Estonian, Latvian, and Lithuanian. From minority languages Sami is a good example beside Basque.

Acknowledgements. This study is used as a part of the EXHADA project by Finnish Academy (#130482).

References

1. Lovins, J.B.: Development of a Stemming Algorithm. Mechanical Translation and Computational Linguistics 11, 23–31 (1968)
2. Harman, D.: How Effective is Suffixing? Journal of the American Society for Information Science 42, 7–15 (1991)
3. Frakes, W.B.: Stemming Algorithms. In: Frakes, W.B., Baeza-Yates, R. (eds.) Information Retrieval, pp. 131–160. Prentice-Hall, Upper Saddle River (1992)
4. Hull, D.: Stemming Algorithms: a Case Study for Detailed Evaluation. Journal of the American Society for Information Science 47, 70–84 (1996)
5. Hollink, V., Kamps, J., Monz, C., de Rijke, M.: Monolingual Document Retrieval for European Languages. Information Retrieval 7, 33–52 (2004)
6. Galvez, C., de Moya-Anegón, F., Herrero-Solana, V.: Term Conflation Methods in Information Retrieval. Non-linguistic and Linguistic Approaches. Journal of Documentation 61, 520–547 (2005)
7. Koskenniemi, K.: Finite State Morphology and Information Retrieval. Natural Language Engineering 2, 331–336 (1996)
8. Loponen, A., Järvelin, K.: A Dictionary- and Corpus-Independent Statistical Lemmatizer for Information Retrieval in Low Resource Languages. In: Agosti, M., Ferro, N., Peters, C., de Rijke, M., Smeaton, A. (eds.) CLEF 2010. LNCS, vol. 6360, pp. 3–14. Springer, Heidelberg (2010)
9. Karlsson, F.: Suomen kielen äänne- ja muotorakenne. WSOY, Helsinki (1983)
10. Kettunen, K., Airio, E., Järvelin, K.: Restricted Inflectional Form Generation in Management of Morphological Keyword Variation. Information Retrieval 10, 415–444 (2007)
11. Kettunen, K., Airio, E.: Is a Morphologically Complex Language Really that Complex in Full-Text Retrieval? In: Salakoski, T., Ginter, F., Pyysalo, S., Pahikkala, T. (eds.) FinTAL 2006. LNCS (LNAI), vol. 4139, pp. 411–422. Springer, Heidelberg (2006)
12. Kettunen, K.: Automatic Generation of Frequent Case Forms of Query Keywords in Text Retrieval. In: Nordström, B., Ranta, A. (eds.) GoTAL 2008. LNCS (LNAI), vol. 5221, pp. 222–236. Springer, Heidelberg (2008)
13. Leturia, I., Gurrutxaga, A., Areta, N., Alegria, I., Ezeiza, A.: Eusbila, a Search Service Designed for the Agglutinative Nature of Basque. In: Lazarinis, F., Vilares, J., Tait, J.I. (eds.) First Workshop on Improving Non English Web Searching (ACM Sigir 2007 Workshop), pp. 47–54 (2007)
14. Leturia, I., Gurrutxaga, A., Areta, N., Pociello, E.: Analysis and Performance of Morphological Query Expansion and Language-filtering Words on Basque Web Searching. In: 6th International Conference on Language Resources and Evaluations (LREC), Marrakech (2008)
15. Paik, J.H., Kettunen, K., Pal, D., Järvelin, K.: Frequent Case Generation in ad hoc Retrieval of Three Indian Languages–Bengali, Gujarati and Marathi. To appear in Proceedings of FIRE 2011 (2012)
16. Kettunen, K., Kunttu, T., Järvelin, K.: To Stem or Lemmatize a Highly Inflectional Language in Probabilistic IR Environment? Journal of Documentation 61, 476–496 (2005)
17. Snowball web site, http://snowball.tartarus.org/
18. Lingsoft, http://www.lingsoft.fi

19. Kettunen, K., Baskaya, F.: Stemming Finnish for Information Retrieval–Comparison of an Old and a New Rule-based Stemmer. In: Vetulani, Z. (ed.) Proceedings of the 5th Language & Technology Conference (LTC 2011), Poznan, pp. 476–480 (2011)

20. Sanderson, M.: Test Collection Based Evaluation of Information Retrieval Systems. Foundations and Trends in Information Retrieval 4, 247–375 (2010)

21. Bar-Ilan, J., Gutman, T.: How do Search Engines Respond to Some Non-English Queries? Journal of Information Science 31, 13–28 (2005)

22. Jansen, B., Spink, A., Sarasevic, T.: Real Life, Real Users, and Real Needs: a Study and Analysis of User Queries on the Web. Information Processing & Management 36, 207–227 (2000)

23. Jansen, B., Spink, A.: An Analysis of Web Searching by European Alltheweb.com Users. Information Processing and Management 41, 361–381 (2005)

Analyzing Query Logs of USPTO Examiners to Identify Useful Query Terms in Patent Documents for Query Expansion in Patent Searching: A Preliminary Study

Wolfgang Tannebaum and Andreas Rauber

Institute of Software Technology and Interactive Systems,
Vienna University of Technology, Austria
{tannebaum,rauber}@ifs.tuwien.ac.at
http://www.ifs.tuwien.ac.at

Abstract. In an attempt to improve retrieval systems for the patent domain, significant efforts are invested to assist researchers in formulating better queries, preferably via automated query generation. Current work on query generation in patent retrieval is mostly based on statistical measures without considering whether these terms are the best choice. To learn from actual queries being posed by experts, we analyze query logs from USPTO patent examiners. Results show that US examiners pick the majority of query terms from the claim section, a large fraction of which, in turn, coincide with the subject feature terms which determine the extent of the protection of the patent right. Considering the lessons learned from evaluating existing search logs will help in improving (semi-) automated query generation.

Keywords: Patent Searching, Query Log Analysis.

1   Introduction

In preparing a patent application or judging the validity of an applied patent based on novelty and inventiveness, an essential task is searching patent databases for related patents that may invalidate the invention. Patent searching is a patent-to-patent associative retrieval task and usually performed by examiners in a patent office and patent searchers in private companies. Just as general information retrieval, patent searching consists of three phases: Query Generation (1), Document Retrieval (2) and Document Reviewing (3). In the query generation process query terms are combined to form a comprehensive query set. In the document retrieval step the patent databases of the national patent offices or commercial operators are searched. Finally, in document reviewing documents are reviewed to select the relevant ones and reiterate the patent search with new queries [6]. Because patent searching is keyword-based and patent applicants are permitted to be their own lexicographers, i.e. they can define their own terminology, the success of patent searching relies on the quality of the query terms used by the patent searchers.

M. Salampasis and B. Larsen (Eds.): IRFC 2012, LNCS 7356, pp. 127–136, 2012.

In this paper our goal is to get a better understanding of which terms of the query documents are defined as the query terms by the literature and the guidelines for patent searching and if these terms are used by the patent examiners for patent searching. Therefore we analyze query logs of USPTO patent examiners and the corresponding query documents. Finally, we take a closer look at acquiring lexical knowledge from patent documents for query expansion in patent searching.

The remainder of the paper is organized as follows. Section 2 provides an introduction into related query generation and expansion experiments in patent searching. Section 3 presents the experiments performed on the dentistry category of the US Patent Application Information Retrieval Register (PAIR). A method to detect query terms in patent documents and an approach to acquire lexical knowledge from patent documents for query expansion in patent searching is presented in Section 4. The conclusions as well as an outlook on future work are provided in Section 5.

2   Related Work

The process of expanding a user`s query with additional related words or phrases can be classified in (1) global, (2) local, or (3) external techniques. Global techniques use corpus-wide statistics. Local techniques analyze for example a number of top-ranked documents retrieved by the query. In external techniques external sources, such as thesauri, are used to find further expansion terms. Furthermore, combinations of several techniques are used to improve query expansion.

Currently, automatic query expansion in patent search is mostly based on global and local techniques. Additional query terms are extracted from the query documents based on statistical measures without considering whether these terms are the best choice. Particularly, additional query terms are extracted using term frequencies (tf) and a combination of term frequencies and inverted document frequencies (tf/idf) [10,11,16,18]. Also whole documents or whole sections of the query documents, like the title, abstract, description or the claim section are used for query generation and expansion. Further approaches use pseudo relevance feedback or citation analysis to expand the query terms from the query documents. Missing terms are discovered from feedback documents or from the cited documents [6,12,19].

Also related work for query expansion in patent searching concerns the expansion of the query terms using external techniques. Therefore current approaches use translation dictionaries, machine translation systems, or parallel corpora or combinations [2,6,9]. A further approach extracts a translation dictionary directly from the patent collection. The approach uses patents granted by the European Patent Office including the claim section in English, French and German to extract translation relations for each language pair. To extract the translations the approach aligns the parallel translations, particularly the whole claim sections [9].

All related approaches to improve query generation and expansion use the whole documents or whole sections of the documents for query generation, query expansion and dictionary learning. The quality of the learned query and expansion terms will not be considered. Learning from actual queries submitted by experts could address this shortcoming of the automatic query generation approaches which only look at statistical information.

3   Query Log Analyses

In several information retrieval applications, such as digital libraries search or web search, query logs are being studied [1,3]. Core of the analysis are basic statistics of the queries, such as the frequencies and the length of the queries, the most co-occurring terms, the click entropy distribution, or the distribution of queries across general topic categories are analyzed [13,14,15,20]. Also the search behavior, particularly the modification of the queries in the session, for example by either adding or deleting query terms, the results of the query sessions and the effectiveness of search systems through query expansion or query suggestion using submitted queries are analyzed [17].

3.1   Experiment Set Up

For our query log analysis experiments we use a collection of query logs of patent examiners freely available from the US Patent and Trademark Office Portal PAIR (Patent Application Information Retrieval), downloaded from (http://www.uspto. gov/). The query logs called "Examiner`s search strategy and results" are published for most patent applications since 2003 by the US Patent and Trademark Office. We collected all patents that are listed under the International Patent Classification (IPC) A61C1 for "Dental machines for boring and cutting" (Dentistry Domain) since 2003 for which the examination procedure is published. We downloaded 162 query documents and the corresponding query logs (346 Logs) from the USPTO Portal PAIR.

Furthermore, for query log term analysis we consider each European patent family document related to the source documents (also referred as query documents) of the query log collection. We retrieved and analyze 107 patent documents applied at the EPO.

3.2   General Statistics

Each query log is a PDF file consisting of a series of queries. Each query has several elements. The Search Query element, showing the query formulated by the patent examiner, is of concern to us here. Further elements are: Reference, Hits, Database(s), Default Operator, Plurals, and Time Stamp. There are several kinds of queries reflected in the search query element. Text queries are used for querying whole documents (fulltext search) or only sections of patent documents, such as the title section (title search). Non-text queries are used for number search, inventor search, or classification search. A further kind of query is the reference query, which is a combination of earlier queries. For query formulation text queries include search operators between the query terms. The types of search operators are (1) Boolean operators, such as "AND or OR", (2) Proximity operators, like "SAME, ADJ(cent), NEAR, or WITH", and (3) Truncation Limiters, such as "$". These operators can be manually added to the text queries by the examiners or automatically computed by the search system. When the operators are added manually, they will be pictured between the query terms in the text query element, else they are indicated by the default operator element.

From the analysis of the layout of the query logs we sample information related to the search query element, particularly to the search queries, for analysis of further general statistics of the query logs. Therefore we concern to the text queries used by the patent examiners for patent searching. Some statistical properties of the text query lengths are listed in Table 1.

Table 1. Query Length Analysis

Text query length	queries	%
text queries	1893	100
unique text queries	**1780**	94,03
1 to 10-term queries	1733	**91,55**
> 10-term queries	47	8,45

We extract 1780 unique text queries from the search query elements of the query log collection. 91,55 % of these have a length between 1 and 10 query terms. Only 8,45 % of the text queries (47 queries) are longer than 10 query terms. As shown in Fig. 1 the maximum number of queries and text queries formulated for a query document are 302 queries and 181 text queries.

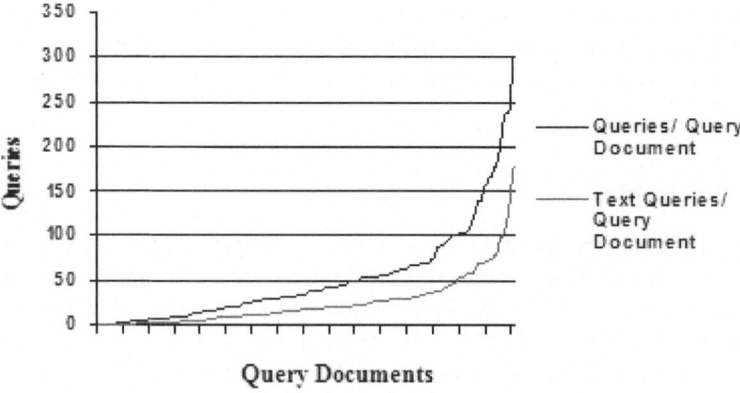

Fig. 1. Number of queries used in the query logs

Some basic statistical properties of the query log collection, particularly the number of unique query terms used for each query document, are listed in Table 2.

Table 2. Query Log Statistics

Query Log Analysis	Mean	Max.	Min.
Query Logs/ Query Document	**2**	17	1
Queries/ Query Document	**53**	302	0
Text Queries/ Query Document	**25**	181	0
Query Terms/ Query Document	**17**	89	0
Query Terms/ Text Query	**4**	54	1

As shown in Table 2 on average two query logs are available for each query document, i.e. the patent examiners searches on average two times for prior art in the examination procedure of a patent application. Each query log includes on average 53 queries for each query document. Therefrom, on average, 25 queries are text queries for fulltext search or patent section search, such as searching the claim sections. We further find that patent examiners formulate queries with an average length of four query terms. On average, 17 unique query terms are used by the examiners to express their information need for each query document.

81,52 % of the search operators between two query terms are manually chosen and appear in the search query element. 19,48 % are default operators and are disclosed by the default operator element. 47,35 % of the manually added operators are proximity operators, such as ADJ(acent), NEAR, SAME, WITH, the remaining operators being the Boolean operators OR (25,47 %) and AND (27,18 %).

We analyze the query terms in view of the part of speech using the CLAWS part of speech tagger for English terms [7]. We analyze the 835 unique query log terms from the query log collection. Table 3 shows the part of speech analysis.

Table 3. Part of Speech Analysis

Part of Speech Characteristics	terms	%
verbs	**129**	**15,66**
unclassified	3	0,36
punctuations	4	0,49
prepositions	3	0,36
ordinal numerals	3	0,36
nouns	**535**	**64,93**
numbers	2	0,24
adverbs	6	0,73
articles	1	0,12
adjectives	**138**	**16,75**
unique query log terms	824	100,00

More than half of the query terms are nouns followed by adjectives and verbs. This shows that patent examiners use predominantly nouns to describe their information need, particularly to compartmentalize the invention into searchable features.

3.3   Query Term Analysis

For studying the terms used by the examiners for formulating text queries, we analyze
the source documents for which the queries are generated, particularly the claim
sections of the query documents. Table 5 shows the analysis of the Query Log Terms
(QLTs) used by the USPTO patent examiners in view of the query documents and the
sections of the query documents.

Table 4. Query Log Term Analysis

Query Log Term Characteristics	avg. terms	%
Unique QLTs/ d	**17**	-
Unique Query Log Terms/ query log collection	835	-
QLTs not in the query document	**4**	17,10
QLTs present in the query document	13	82.90
QLTs present only in the Description of d	2	14.98
QLTs present in the Claim Section of d	**11**	**85.02**

The majority of query terms (13 of 17) that the patent examiners use come from
the patent application (constituting the query document d). On average, the USPTO
patent examiners selected 82.90 % of the query terms from the query document. The
other 17.10 % of the QLTs are expansion terms which do not appear in the query
document. This means, that the examiners expand on average the 13 QLTs from the
query document with further four expansion terms by brainstorming. The examiners
select, on average, 85.02 % of the QLTs from the claim sections. Only 14.98 % of the
QLTs are from the description section. These terms do not appear in the claims.
Otherwise, QLTs which appear in the claim section can also occur in the description
section.

Claim Term Analysis

We further study which terms from the claim sections are used for query generation
by the USPTO patent examiners.

The claims must define the matter for which protection is sought in terms of the
technical features of the invention. In several patent conventions, such as the
European, patent applications containing drawings with reference signs linking
the claims to the drawings the subject/ technical features *SFTs* specified in the claims
are preferably followed by reference signs relating to these features. The reference
signs are placed in parentheses [4,5].

Hence, for the claim term analysis we consider each European patent family
document related to a US query document (member of the patent family) of the query
document collection including the claim section in English and containing reference
signs next to the *SFTs*. We retrieved and analyzed 107 patent family documents
applied at the EPO. Finally, we compare the query terms from the claim sections of

the US query documents with the neighbor terms, particularly the *SFTs*, of the reference signs in the European patent family documents.

We find that on average 30,40 % of the claim terms used by the examiners for query formulation co-occur a reference sign. Furthermore, we figure out that with 17,66 % of the claim terms appear in a phrase beside a reference sign. Therefore we compared the query terms with the terms from the patent family documents occurring on a second position next to the reference signs. Totally, on average 48,06% of the query terms selected from the claim section appear next to the reference signs.

Furthermore, we study, if the other 51.95 % of the claim terms used by the examiners for query formulation are *SFTs* in the specific patent domain. We find that 73,17 % of these terms which do not appear next to a reference sign in the query document appear next to a reference sign in one or more patents filed in the respective domain. Considering phrases 87,80 % of the terms appear in the domain next to reference signs.

Expansion Term Analysis

We also analyze the *QLTs* which do not appear in the query documents, also called expansion terms *ETs*. Therefore we queried the patent domain using the *ETs* from the query log collection (349 unique *ETs*). Table 6 shows the results of the analysis of the expansion terms.

Table 5. Expansion Term Analysis

Expansion Term Characteristics	terms	%
Expansion vocabulary:	**349**	100,00
ETs appear in the domain:	291	**83,38**
ETs not in the domain:	58	16,62
ETs occur next to a reference sign:	204	**70,10**
ETs occur not next to a reference sign:	145	29,90

We find that 83,38 % of the used vocabulary for query expansion appear in the specific patent domain. Therefrom 70,10 % of the terms co-occur with a reference sign. Finally, we learn from the Claim Term Analysis and from the Expansion Term Analysis that US Examiners select predominantly the *SFTs* from the query document as proposed by the literature for patent searching and brainstorm a small number of additional terms, which we also identified as domain specific *SFTs*.

4   Acquiring Lexical Knowledge from Patent Documents

In traditional patent searching, to scope a search, the first step is to compartmentalize the invention into searchable features to create a so called invention diagram. The invention diagram serves as a template to brainstorm synonyms and equivalents to the co-occurring features of the invention to create the expanded invention diagram [8].

Figure 2 shows an example for such an expanded diagram.

Features	Expansion Terms
control module	control, module, terminal, computer, station, …
display	screen, keypad, touchpad, Interface, …
transmitter	radio, infrared, ultrasonic, transmit, send, signal,
receiver	radio, infrared, ultrasonic, receive, reception, …

Fig. 2. Expanded invention diagram

The first column includes the features of the invention selected from the source document or an invention report. The second column provides the corresponding expansion terms to the features of the invention. Hence, the requirements to an expansion tool for query expansion in patent searching are to suggest synonym and co-occurring expansion terms. A lexical ontology might meet these requirements.

4.1   Experiment Set Up

For our experiments to detect *SFTs* in patent documents and to acquire lexical knowledge from patent documents for query expansion in patent searching we use a collection of granted European patents which have to be filed in one of the official languages of the European Patent Office (EPO), namely English, French or German, and include also, if granted by the EPO, the claim section in the other two official languages for ontology enrichment. We collect all granted European patent (453 documents) for the specific IPC class A61C1 from the European Publication Server (data.epo.org). We downloaded 276 documents available in fulltext. From each document we extract the claim sections, which will be published in French, English and German, for our experiments.

4.2   SFT and Lexical Knowledge Detection

For *SFT* detection in the claim sections of the patents we use the reference signs appearing next to the *SFTs*, as proposed in the claim term analysis section. We extract the *SFTs* by co-word analysis of the reference signs.

Our approach to acquire lexical knowledge works as follows: First, we learn translation relations using word co-occurrences. Bigrams from the claim sections including the same reference sign are assembled to translation relations. Therefore, we first filter all generated bigrams of the claim collections in the form "*S R*", where *S* is a claim term, particularly a *SFT* and *R* a reference sign, then we filter the retrieved bigrams from step 1, where bigrams with the same reference sign occur more than two times. We then generate translation relations based on the bigrams including the same reference signs. Based on these learned relations including equal concepts we learn synonym relations. Table 6 shows the semantic relations learned from the European patents.

Table 6. Learned Semantics

Lexical Relations	Code	semantic relations	concepts
English/ German	EG	1443	1370
English/ French	EF	1158	932
German/ French	GF	1264	1401
English/ English	EE	305	305
German/ German	GG	193	271
French/ French	FF	120	167
$\sum$ unique relations and concepts		**4483**	**2090**

We extract from the document collection, particularly from the claim sections of the granted European patents, 4483 unique semantic relations including 2090 unique concepts for the specific IPC class A61C1. Further, we learn 305 English, 193 German, and 120 French synonym relations based on same translation terms in the learned translation relations. We do not learn semantics which conflict, particularly semantics which include the same concepts and appear as synonym and translation relations. For example, for the concepts "drill" and "bit" the lexical ontology provides only a synonym relation.

Compared to related work, which aligns the whole claim sections of the European patents for translation dictionary learning, we limit the lexical ontology learning approach on claim terms, particularly on the subject feature terms, which are identified as query terms by the patent literature and which are used by the examiners for query generation. Further we learn lexical knowledge based on the learned translation relations directly from the patent document collection for query expansion in patent searching.

5   Conclusions and Future Work

In this paper we studied the literature for traditional query generation and query expansion in patent searching and analyzed a collection of query logs of USPTO patent examiners to identify useful query terms in patent documents for query expansion in patent searching.

The results show that patent examiners formulate queries with an average length of four query terms. The majority of the query terms used for document retrieval come from the patent application. Therefrom the US Examiners select predominantly the *SFTs* as proposed by the literature for patent searching and brainstorm a small number of additional terms, which we also identified as domain specific *SFTs*.

Furthermore, we presented a method to detect *SFTs* in patent documents and an approach to acquire lexical knowledge from the documents for query expansion in patent searching. Lexical knowledge could be automatically extracted from the query documents without using semantics from external sources, such as lexica or databases, for relation finding. Further, we learned a lexical ontology based on the *SFTs*, which are identified as the query terms by the patent literature and which are used by the examiners for query generation.

Future work will focus on the evaluation of the lexical knowledge. We want to use a set of query logs, particularly text queries of search logs, to evaluate the learned semantic relations in query expansion. Further we want to evaluate the performance of the generated lexical ontology based on patent searches based on Recall and Precision in terms of the success of query expansion.

References

1. Amitay, E., Broder, A.: Introduction to special issue on query log analysis: Technology and Ethics. ACM Trans. Web 2, Article 18 (2008)
2. Ballesteros, L., Croft, W.: Phrasal translation and query expansion techniques for cross-language information retrieval. In: Proc. of SIGIR 1997, Philadelphia, Pennsylvania, USA, pp. 84–91 (1997)
3. Clough, P., Berendt, B.: Report on the Treble CLEF query log analysis workshop 2009. SIGIR Forum 43, 71–77 (2009)
4. European Patent Office. Guidelines for Examination in the European Patent Office (2010)
5. European Patent Office. How to get a European patent – Guide for Applicants Part 1, 13th edn. (2010)
6. Fujita, S.: Technology survey and invalidity search: An comparative study of different tasks for Japanese patent document retrieval. Information Processing and Management, An International Journal, 1154–1172 (2007)
7. Garside, R., Smith, N.: A hybrid grammatical tagger: CLAWS4. In: Garside, R., Leech, G., McEnery, A. (eds.) Corpus Annotation: Linguistic Information from Computer Text Corpora. Longman, London, pp. 102–121 (1997)
8. Hunt, D., Nyugen, L., Rodgers, M.: Patent Searching: Tools & Techniques. John Wiley & Sons, Inc. (2007)
9. Jochim, C., Lioma, C., Schütze, H., Koch, S., Ertl, T.: Preliminary study into query translation for patent retrieval. In: Proc. of PaIR 2011, Toronto, Canada, pp. 57–66 (2010)
10. Konishi, K.: Query terms extraction form Patent Documents for invalidity search. In: Proc. of NTCIR 2005: NTCIR-5 Workshop Meeting, Tokyo, Japan (2005)
11. Magdy, W., Jones, G.J.F.: A Study of Query Expansion Methods for Patent Retrieval. In: Proc. of PaIR 2011, Glasgow, Scotland, pp. 19–24 (2011)
12. Magdy, W., Leveling, J., Jones, G.J.F.: Exploring Structured Documents and Query Formulation Techniques for Patent Retrieval. In: Peters, C., Di Nunzio, G.M., Kurimo, M., Mostefa, D., Penas, A., Roda, G. (eds.) CLEF 2009. LNCS, vol. 6241, pp. 410–417. Springer, Heidelberg (2010)
13. Murray, G.C., Teevan, J.: Query log analysis: Social and technological challenges. SIGIR Forum 41, 112–120 (2007)
14. Silvestri, F.: Mining Query Logs: Turning Search Usage Data into Knowledge. Foundations and Trends in Information Retrieval 4(1-2), 1–174 (2010)
15. Silverstein, C., Marais, H., Henzinger, M., Moricz, M.: Analysis of a very large web search engine query log. SIGIR Forum 33, 6–12 (1999)
16. Torres, S., Hiemstra, D., Serdyukov, P.: Query log analysis in the context of information retrieval for children. In: Proc. of SIGIR 2010, Geneva, Switzerland, pp. 847–848 (2010)
17. Tyler, S., Teevan, J.: Large scale query log analysis of re-finding. In: Proc. of WSDM 2010, New York, USA, pp. 191–200 (2010)
18. Xue, X., Croft, W.: Automatic query generation for patent search. In: Proc. of CIKM 2009, Hong Kong, China, pp. 2037–2040 (2009)
19. Xue, X., Croft, W.: Transforming patents into prior-art queries. In: Proc. of SIGIR 2009, Boston, USA, pp. 808–880 (2009)
20. Zhang, Z., Yang, M., Li, S., Qi, H., Song, C.: Sogou Query Log Analysis: A Case Study for Collaborative Recommendation or Personalized IR. In: Proc. of IALP 2009, Singapore, pp. 304–307 (2009)

Discovering Relevant Features for Effective Query Formulation

Luepol Pipanmaekaporn and Yuefeng Li

School of Electrical Engineering and Computer Science,
Queensland University of Technology, Brisbane, Australia
n7047282@student.qut.edu.au, y2.li@qut.edu.au

Abstract. The quality of discovered features in relevance feedback (RF) is the key issue for effective search query. Most existing feedback methods do not carefully address the issue of selecting features for noise reduction. As a result, exracted noisy features can easily contribute to undesirable effectiveness. In this paper, we propose a novel feature extraction method for query formulation. This method first extract term association patterns in RF as knowledge for feature extraction. Negative RF is then used to improve the quality of the discovered knowledge. A novel information filtering (IF) model is developed to evaluate the proposed method. The experimental results conducted on Reuters Corpus Volume 1 and TREC topics confirm that the proposed model achieved encouraging performance compared to state-of-the-art IF models.

Keywords: Query Formulation, Relevance Feedback, Information Filtering, Pattern Mining.

1 Introduction

Relevance feedback (RF) has been demonstrated very effective for improving retrieval accuracy. RF allows users to describe their information needs by giving relevant (positive) and non-relevant (negative) documents. The basic idea is to utilise the relevant information for describing the intent of query (query refinement) [7,4] or describing user's search intent (user profile) [25,32].

The process of query refinement or query expansion modifies the original keyword query submitted by the user to the search engine. The new query is then used as input to the search engine's ranking algorithm in order to improve the retrieval results. Many retrieval models such as BM25 [22], Rocchio [20], and probabilistic models [21] make use of the RF to find new query terms that will be added to the original query. Some retrieval models such as language models for IR [7] and Markov Random Field [17] use pseudo-relevance feedback especially when there were no relevance judgements available. In order to improve query refinement, several methods focus on identifying good expansion terms in RF. For example, in [4] a supervised learning method was used to select expansion terms in documents associated with the original query. A number of works play attention to find good search terms from the most relevant

M. Salampasis and B. Larsen (Eds.): IRFC 2012, LNCS 7356, pp. 137–151, 2012.
© Springer-Verlag Berlin Heidelberg 2012

areas of document [16,11] as well as co-occurrence features [20,30]. Currently, some researchers demonstrated that phrases are useful for query expansion [6,1]. However, most existing methods for query expansion focus on extracting query terms from pseudo-relevance feedback. As a result, the quality of selected terms strongly relies on the quality of the top-ranked documents [13].

The problem of formulating queries (or user profiles) can be treated as a learning problem. Given a set of training examples, some supervised learning methods can be applied to learn a model of user's interests (query) according to the training samples. For example, both Rocchio and SVM methods are the two popular methods in this area [25,15]. Some probabilistic methods [33,18] have been proposed to estimate the importance of terms with a training corpus. However, these methods have the well-known problem that is the high-dimensionality of text. Many irrelevant terms selected unintentionally can be harmful to the retrieval effectiveness. The challenging issue is how to select a limited number of relevant terms from a large pool of candidate terms for each specific query.

Recently, some promising pattern-based methods have been proposed to solve this challenge [31,34]. In [31], term association patterns were extracted from a training set of documents to reduce a number of noisy terms with respect to a certain frequency. After that, the extracted patterns were used to generate a weighted vector of terms as a query (or user profile) for the training set. Some of the works [34,14] make use of negative feedback to improve the quality of query terms. According to the experimental results, the pattern-based methods achieved excellent performance compared to state-of-the-art term-based methods such as BM25 and Rocchio. Nevertheless, the challenge issue is how to guarantee the quality of discovered patterns for extracting high-quality term. Some discovered patterns may include noisy terms that cause producing unsatisfying search results.

Motivated by the above challenges, we focus on addressing the issue of extracting high-quality features in RF for satisfying users' long-term specific interests, called *specific features*. We proposed a novel pattern mining method for query formulation. Given a training set of documents, this method first discovers association patterns from positive documents for noise reduction. Negative documents are then mined to deal with irrelevant terms that cause producing poor queries. We evaluated the proposed method by developing a novel information filtering (IF) model. The experimental results conducted on Reuters Corpus Volume 1 and TREC topics demonstrate that the proposed model achieved encouraging performance as compared to state-of-the-art IF models including pattern-based and term-based models.

2   Related Work

Relevance Feedback (RF) has been widely used in different retrieval models: Vector Space Model (VSM) [24,20], BM25 [24], probabilistic models [21], language models for IR [7] , and Markov Random Field [17]. There are two main approaches for utilising RF: query refinement and query formulation. Since our work focuses on query formulation, we will review the related studies in this area.

2.1   Query Formulation

The process of query formulation learns a model of user's interest via a training corpus. The user's interest model (i.e., a user profile) can be used to score documents or to make a binary decision (i.e., relevant or non-relevant). Many methods have been proposed as a basis for constructing the user profile model. For example, user profiles can be constructed according to a training set of documents by using some supervised learning methods such as Rocchio and SVM [25] as well as partially supervised learning method [15]. Some probabilistic methods [33,18] have been proposed to estimate the importance (weights) of terms in a training corpus. Nevertheless, these methods have the well-known problem that is the high-dimensionality of text.

In order to improve query formulation, many approaches focus on extracting good terms (features) for a representation of user profile. In [8], Latent Semantic Indexing (LSI) was used for selecting search terms from true relevant documents according to a semantic dimension space. In [12], both linguistic features and statistical features were utilised to select good search terms. Some methods for selecting expansion terms in pseudo relevance feedback can be applied to query formulation [16,11].

Our work differs from the existing works in that we focus on extracting high-quality search terms from RF using data mining techniques.

2.2   Phrase Extraction

Our work is also related to phrase extraction. Some retrieval models and text mining models make use of various phrases [19,5] as text features since they have more precise than single words. However, many experiments do not support the effective use of phrases due to the feature sparseness in text and too many phrases generated without meaning [29,26]. Currently, some IR researchers found that combining phrases and terms could significantly improve the retrieval effectiveness [1,19].

In the presence of these setbacks, closed sequential patterns used in data mining community have turned out to be a promising alternative to phrases [32,9] because they enjoy good statistical properties like terms, and have a reasonable number of generated patterns. Nevertheless, the key issue is how to deal with the closed patterns to improve the retrieval effectiveness. To effectively use closed patterns in text, a deploying method in [31] has been proposed to compose all closed patterns of a category into a vector that included a set of terms and a term-weight distribution. The pattern deploying method has shown encouraging improvements on effectiveness in comparing with traditional IR models. In [34], the authors make use of mining negative documents to enhance the quality of search terms by revising weights of query terms.

Our work differs from the existing works in the following points: 1) Our goal is to discover useful patterns from positive and negative documents for extracting good search terms and 2) we focus on removing noisy terms that cause producing poor queries.

3   Basic Definitions

In this section, we give the brief definitions of frequent patterns, sequential patterns, closed patterns in text, and the pattern deploying method. In this paper, all documents are divided into paragraphs. So, a given document d yields a set of paragraphs $PS(d)$. Let D be relevance feedback, including a set of positive (relevant) documents, D^+, and a set of negative (irrelevant) ones, D^-.

3.1   Frequent Patterns and Closed Patterns

Let $T = \{t_1, t_2, \ldots, t_m\}$ be a set of terms which are extracted from positive documents D^+. Given a *termset* X be a set of terms in a document d, i.e., $X \subseteq T$, $coverset(X) = \{d_p | d_p \in PS(d), X \subseteq d_p\}$. Its absolute support

$$Sup_a(X) = |coverset(X)| \tag{1}$$

and its relative support

$$Sup_r(X) = \frac{|coverset(X)|}{|PS(d)|} \tag{2}$$

A termset X is called *frequent patterns* if its sup_a or $(Sup_r) \geq min\_sup$, a minimum support. Given a set of paragraphs $Y \subseteq PS(d)$, we can define its termset, which satisfies $termset(Y) = \{t | \forall dp \in Y \Rightarrow t \in dp\}$.

Let $Cls(X) = termset(coverset(X))$ be the closure of X. We call X *closed* if and only if $X = Cls(X)$. Let X be a closed pattern. We have $Sup_a(X_1) < Sup_a(X_2)$ for all pattern $X_1 \supset X_2$.

3.2   Closed Sequential Patterns

A sequence $X = \langle t_1, \ldots, t_r \rangle$ $(t_i \in T)$ is an ordered list of terms. Let a sequence $s_1 = \langle x_1, \ldots, x_i \rangle$ is a sub-sequence of another sequence $s_2 = \langle y_1, \ldots, y_j \rangle$, denoted by $s_1 \sqsubseteq s_2$, iff $\exists j_1, \ldots, j_i$ such that $1 \leq j_1 < j_2 \ldots < j_i \leq j$ and $x_1 = y_{j_1}, x_2 = y_{j_2}, \ldots, x_i = y_{j_i}$. Given $s_1 \sqsubseteq s_2$, we usually say s_1 is a *sub-pattern* of s_2, and s_2 is a *super-pattern* of s_1.

A sequence X is called a *sequential pattern* if its absolute or relative support $\geq min\_sup$. A frequent sequential pattern X is call *closed* if not $\exists$ any super-pattern X_1 of X such that $Sup_a(X_1) = Sup_a(X)$.

To improve the efficiency of pattern mining, an algorithm $SPMining(D^+, min\_sup)$ have proposed in [32] to efficiently extract closed sequential patterns for all positive documents D^+ based on a given min_sup (we do not repeat the algorithm due to the limited length of this paper).

3.3   The Pattern Deploying Method

A pattern deploying method (PDM) has been proposed in [31]. The idea of the PDM is to use closed sequential patterns (hereafter patterns) discovered in the training documents for evaluating supports (weights) of each term for the training set. The PDM adopts the $SPMining$ algorithm [32] to discover patterns in positive documents. For each document $d_i \in D^+$, this algorithm extracts a set of patterns SP_i, based on a given min_sup.

Let $SP_1, SP_2, \ldots, SP_{|D^+|}$ be the sets of discovered patterns for all document $d_i \in D^+$, where $i = 1, \ldots, |D^+|$. For a given term $t \in D^+$, its support (weight) in the patterns is assigned by:

$$support(t, D^+) = \sum_{i=1}^{n} \sum_{t \in p \subseteq SP_i} \frac{\sup_r(p, d_i)}{|p|} \tag{3}$$

where $|p|$ is the number of terms in p. These terms were used to form a query vector for information filtering.

4 Our Approach

The proposed approach consists of two parts: *pattern refinement* and *Query extraction*. We assume that a set of patterns for all positive documents D^+, denoted F^+, has already been mined. We called *positive patterns*.

4.1 Pattern Refinement

In this section, we describe the method for refining positive patterns, aiming to improve the quality of patterns.

4.1.1 Mining Negative Relevance Feedback

Negative Relevance Feedback (NRF) contains non-relevant information. NRF is useful for estimating the boundary of positive documents. In other words, NRF allows to capture the differences between relevance and non-relevance. In this work, our idea is to utilise the negative feedback to deal with the distracting relevant information represented by positive patterns and to enhance the quality of terms extracted. Figure 1 shows relations between patterns from positive and negative documents.

According to Figure 1, some positive and negative patterns may contain overlapping features (terms) (i.e., unshaded triangles). These terms may be considered as *noise* since they appear in both positive and negative documents. Although using pattern mining allows to reduce some noisy terms w.r.t. frequency, many *noisy frequent terms* (shaded triangles) are still available). For example, in Figure 1 positive pattern C contains one noisy frequent term and pattern E that contains all the noisy terms. With the negative feedback information, we can characterize noisy patterns in true relevant documents into one of the follow categories:

Definition 1 (Meaningless patterns). *A positive pattern p is a meaningless pattern if it is a subset of negative patterns.*

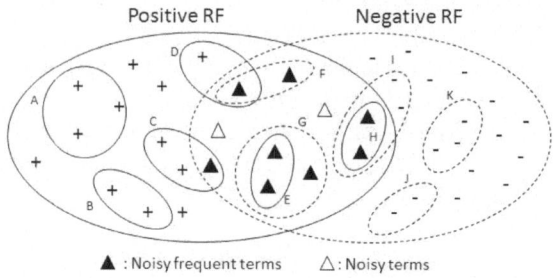

Fig. 1. Relations between positive and negative patterns

For example, both positive patterns E and H in Figure 1 are meaningless since they are a subset of negative pattern G and I (dashed-line ovals) respectively. This type of noisy patterns should be eliminated since they frequently appear in the context of non-relevant information.

Definition 2 (Weak patterns). *A positive pattern p is a weak pattern if it partially overlaps with negative patterns.*

For example, positive patterns C and H in Figure 1 are examples of the weak positive patterns. This kind of noisy patterns presents a significant challenge to be managed. Eliminating them may cause losing some relevant information. Rather than removing them, we deal with the weak patterns by adding some negative patterns related to them to balance weights of noisy terms that cannot be eliminated at the level of pattern.

4.1.2 Selecting Negative Documents

A collection of negative documents contains non-relevant information that might be quite diverse. This is because the negative documents may come from unrelated topics. In order to effectively mine the negative samples, we propose to select some *offenders* in negative documents, where an offender is a negative document that tends to be identified falsely as relevant by the system.

Given T^+ be a set of frequent terms extracted from all positive patterns in F^+, we use cosine similarity to assign weights to all negative documents $nd \in D^-$ for identifying the offenders as follows:

$$Sim(nd, T^+) = \sum_{t_j \in nd \cap T^+} tf_j \left/ \sqrt{\sum_{t_k \in nd \cap T^+} tf_k^2} \right. \tag{4}$$

where nd is a negative document in D^- and tf_j is the frequency of term t_j in the document. The document nd is called *offender* if its weight is positive (i.e., $Sim(nd, T^+) > 0$). In this paper, only the top-n offenders is used to replace a whole collection of negative documents, where $n = |D^+|$.

After the set of negative offenders D^-_{off} was obtained, $SPMining$ algorithm is called to to mine a set of patterns F^- in the selected negative samples. We perform to refine the positive patterns as follows:

1) For each pattern np in the collection F^- do
   Eliminate all positive patterns p in F^+ that are a subset of pattern np, i.e.,
   $F^+ = F^+ - \{p|p \in F^+, \exists np \in F^- \Rightarrow p \subseteq np\}$.
2) For each pattern q in the collection F^+ do
   Select negative patterns np in F^- that overlap with the pattern q, i.e.,
   $F^- = F^- - \{np|np \in F^-, \exists q \in F^+ \Rightarrow q \cap np \neq \emptyset\}$.

4.2   Feature Extraction

In this section, we describe the feature extraction method using patterns. We first extract candidate terms from positive documents D^+.

For each document $t \in D^+$, the weight of the term t is calculated according to the following function:

$$w(t, D^+) = \frac{\sum_{i \in D^+} \lambda_i(t) - \sum_{j \in D^-_{off}} \lambda_j(t)}{|D^+| + |D^-_{off}|} \qquad (5)$$

where $\lambda_i(t)$ and $\lambda_j(t)$ are weight functions that return the weight of term t associated with the positive document d_i and negative document d_j respectively. For each positive document d_i, we define the weight function as follows:

$$\lambda_i(t) = \sum_{t \in p \subseteq SP_i \cap F^+} \frac{|p|}{|p| + 1} \times sup_r(p, d_i) \qquad (6)$$

where $|p|$ is the number of terms in the pattern p, SP_i is a set of patterns discovered in the document d_i, and $Sup_r(p, d_i)$ is the relative support of pattern p in the document d_i.

According to the above function (Eq.(5)), the weight of term t is determined by averaging weights assigned by positive and negative documents (i.e., offenders). For each positive document, the weight of term is evaluated according its appearances in the patterns contained in the document and F^+ (See Eq. 6). In order to obtain the term weight, we combine two components of pattern. The first component (i.e., $\frac{|p|}{|p|+1}$) reflects the importance of pattern. We simply calculate the pattern weight by normalizing the length of pattern. This ensures that the term extracted from large patterns carrying more specific information of the document is more important than the one extracted from shorter patterns that usually carry general information. The second component of the support function is the frequency of pattern (i.e., $sup_r(p, d_i)$) that reflects the reliability of the pattern. A pattern with low support is unreliable.

We also make use of negative documents to evaluate the weight of term in order to suppress the effect of noisy terms that cannot be removed at the pattern, i.e., $\lambda_j(t) = \sum_{t \in q \subseteq SP_j \cap F^-} \frac{|q|}{|q|+1} \times sup_r(q, d_j)$. Finally, terms with positive weight (i.e.,$w(t, D^+) > 0$) are selected to formulate the query vector Ω. We normalise the vector Ω with the norm of the query $||\Omega|| = (\Omega.\Omega)^{\frac{1}{2}}$ where . is the dot product.

A document matching function r is simply built for the use of the query in information filtering. Given incoming documents d, the relevance score of the document is computed by:

$$r(d) = \sum_{t \in \Omega} w_\Omega(t) \times \tau(t, d) \tag{7}$$

where $w_\Omega(t)$ is the weight of query term t in Ω and $\tau(t, d) = 1$ if $t \in d$; otherwise $\tau(t, d) = 0$.

A high value assigned to the document can imply that the document is *highly relevant*.

5   Evaluation and Discussion

5.1   Experimental Dataset

Reuters Corpus Volume 1 (RCV1) data collection [23] is used to evaluate the proposed model, denoted *QFM* (Query Formulation model). This corpus consists of all and only English language stories proposed by Reuter's journalists between 1996-08-20 to 1997-08-19, a total of 806,791 documents that cover very large topics and information. TREC (2002) has developed and provided 50 assessor topics for the TREC filtering track, aiming to building a robust filtering system [28]. The TREC topics were developed by human assessors of the National Institute of Standards and Technology (NIST), called *assessor* topics. According to [3], using at least 50 queries to evaluate a system is stable and sufficient for a good experiment. Thus, this research uses RCV1 and the 50 assessor topics to evaluate the proposed model.

5.2   Data Preprocessing and Measures

For each assessor topic, its data collection is split into two sets: a training set and a test set. All documents are marked in XML and some meta-data information. We remove all the meta-data information and perform a common basic text processing for all documents, including stop-words removal according to a given stop-words list ans stemming terms using Poster Stemmer.

The effectiveness is measured by five standard IR metrics: The precision at the top 20 ranked documents ($top - 20$), $F_{\beta=1}$ measure, Mean Average Precision (MAP), the break-even point (b/p), and Interpolated Average Precision (IAP) on $11-$points. Precision (p), Recall (r are calculated by the following functions: $p = \frac{TP}{TP+FP}$ and $r = \frac{TP}{TP+FN}$, where TP is the number of documents the system correctly identifies as positives; FP is the number of documents the system falsely identifies as positives; FN is the number of relevant documents the system fails to identify.

5.3   Baseline Models and Settings

We grouped baseline models into two main categories. The first category includes a number of patterns-based methods for Information Filtering (IF):

- **PTM** [32]: This method discovers all closed sequential patterns from a collection of positive documents as features for a user profile. Weights are assigned to each pattern based on its appearance in the positive and negative documents.
- **PDM** [31]: This method uses closed sequential patterns discovered in positive documents for formulating a vector of terms as a user profile. Instead of normal term evaluations, weights of query terms are evaluated based on their appearance in the document patterns (See the details in Section 2).
- **IPE** [34]: A method for tuning a user profile [31], using negative feedback has been proposed. It extracts features from non-relevant documents to revise the weight of query term according to their appearances in the negative documents. According to the experimental results, IPE is much effective than popular term-based IF methods such as BM25 and Rocchio. It also outperforms PDM in average performance on all RCV1 topics.

For all the pattern-based methods, the minimum support threshold (min_sup) was set to 0.2 (20% of the number of paragraphs in a document) for this experiment. This is because we found that it was the best value (also used in the experiments [32,31,34]).

The second category of our baseline methods includes popular term-based methods in IR:

- **Rocchio** [10]: This method generates a Centroid for representing user profiles by extracting terms from positive documents and performing to revise weights of the terms with negative documents. The centroid c of a topic can be generated as follows:

$$c = \alpha \frac{1}{|D^+|} \sum_{\vec{d} \in D+} \frac{\vec{d}}{||\vec{d}||} - \beta \frac{1}{|D^-|} \sum_{\vec{d} \in D-} \frac{\vec{d}}{||\vec{d}||} \qquad (8)$$

where $||\vec{d}||$ be normalized vector for document d. α and β be a control parameter for the effect of relevant and non-relevant data respectively. According to [10,2], there are two recommendations for setting the two parameters: $\alpha = 16$ and $\beta = 4$; and $\alpha = \beta = 1.0$. We have tested both accommodations on assessor topics and found the latter recommendation was the best one. Therefore, we let $\alpha = \beta = 1.0$.

- **BM25** [22]: a well-known ranking function for scoring matching documents according to a given search query has been considered the state-of-the-art baseline in IR. We use the BM25 function to estimate a weight of query term t extracted from positive document as follows:

$$W(t) = \frac{tf \cdot (k_1 + 1)}{k_1 \cdot ((1 - b) + b\frac{DL}{AVDL}) + tf} \cdot \log \frac{\frac{(r+0.5)}{(n-r+0.5)}}{\frac{(R-r+0.5)}{(N-n-R+r+0.5)}} \qquad (9)$$

where N is the total number of training documents; R is the number of positive documents; n is the number of documents which contain term t; r is the number of positive documents which contain term t; tf is the term frequency; DL and $AVDL$ are the document length and average document length, respectively; and k_1 and b are the experimental parameters. We use the BM25 with the parameters tuned in [34] (i.e.,$k_1 = 1.2$ and $b = 0.75$).

- **Support Vector Machine (SVM)** : the linear SVM has been proven very effective for text categorization and filtering [27]. We would compare it with QFM. However, most existing SVMs are designed for making a binary decision rather than ranking documents. We choose the linear SVM modified for document retrieval in [34]. The SVM only uses term-based features extracted from training documents. We describe the details as following:

There are two classes: $y_i \in \{-1, 1\}$ where $+1$ is assigned to a document if it is relevant; otherwise it is assigned with -1 and there are N labelled training examples: $(d_i, y_i), \ldots, (d_N, y_N)$, $d_i \in \mathbb{R}^n$ where n is the dimensionality of the vector. Given a function $h(d) =< w \cdot d > +b$ where b is the bias, $h(d) = +1$ if $< w.d > +b \geq 0$; otherwise $h(d) = -1$, and $< w \cdot d >$ is the dot product of an optimal weight vector w and the document vector d. To find the optimal weight vector w for the training set, we perform the following function: $w = \sum_{i=1}^{N} y_i \alpha_i d_i$ subject to $\sum_{i=1}^{l} \alpha_i y_i = 0$ and $\alpha_i \geq 0$, where α_i is the weight of the sample d_i. For the purpose of ranking, b can be ignored and all training documents are important equally. We hence assign the same α_i value (i.e., 1) to each positive document first, and then determine the same α_i (i.e., α') value to each negative document. Thus, the optimal weight vector w can be determined as following function:

$$w = \left(\sum_{d_i \in D^+} d_i \right) + \left(\sum_{d_j \in D^-} d_j \alpha' \right) \tag{10}$$

In order to score a matching document d, the ranking function $S(d) = d \cdot w$ is performed. A high positive value assigned to the document d can imply that the document tends to be *highly relevant*.

For all the term-based methods, we perform to choose top$-k$ terms from positive documents for each topic query based on $tf \times idf$. The average number of document terms for each topic is 629.28. The results have shown in Table 1.

Table 1. Mean Average Precision w.r.t. Top$-k$ terms

Methods	Top-k terms								
	50	100	150	200	250	300	400	500	ALL
Rocchio	0.411	0.423	0.428	0.434	0.436	**0.440**	0.438	0.436	0.433
SVM	0.401	0.414	0.423	0.430	**0.432**	0.430	0.426	0.428	0.424
BM25	0.395	0.406	0.411	0.412	**0.418**	0.415	0.414	0.410	0.406

5.4 Minimum Support Thresholds

Minimum support threshold *min_sup* can strongly affect the performance of pattern-based models. A high minimum support may miss important patterns while a small one may result in discovering a large number of noisy patterns. In this section, we examine the effect of minimum support thresholds associated with Mean Average Precision(MAP) on all assessor topics. Figure 2 illustrates the results of varying the minimum support values.

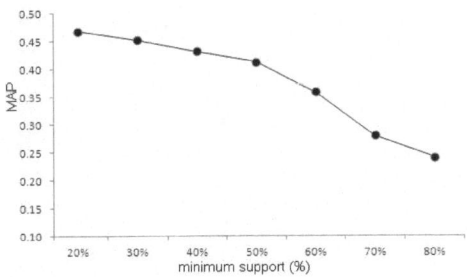

Fig. 2. MAP w.r.t. Minimum support thresholds

According to Figure 2, we found that QFM achieved the best MAP performance at *min_sup* = 20%. Increasing the minimum support threshold leads to performance drop. Please note that setting *min_sup* = 10% results in the *SPMining* algorithm that cannot run completely.

Table 2. Comparison results with all methods on the First 50 RVC1 Topics

Methods	*top-20*	*MAP*	*b/p*	$F_{\beta=1}$	$Avg_{|Q|}$ (%Chg)
QFM	**0.545**	**0.473**	**0.457**	**0.464**	102.70 (−83.52%)
PDM	0.496	0.444	0.430	0.439	156.90 (−75.07%)
IPE	0.493	0.441	0.429	0.440	156.90 (−75.07%)
PTM (Closed Seq. Ptrns)	0.406	0.364	0.353	0.390	199.16 (−68.35%)
PTM (Seq. Ptrns)	0.401	0.361	0.343	0.385	398.18 (−36.72%)
Rocchio with $\alpha = 1.0$ and $\beta = 1.0$	0.491	0.440	0.411	0.435	300.00 (−52.32%)
SVM with $b = 0$	0.483	0.432	0.402	0.4317	250.00 (−60.27%)
BM25 with $k_1 = 1.0$ and $b = 0.75$	0.469	0.418	0.420	0.387	250.00 (−60.27%)

5.5 Comparison to Pattern-Based Methods

Table 2 compares the performance of all pattern-based methods and the proposed method, QFM. The most important findings revealed in this table are that PTM with closed sequential patterns performs better than PTM with non-closed ones

Table 3. $p-$values for the baseline methods comparing with QFM in the assessor topics

Model	top-20	MAP	b/p	$F_{\beta=1}$
PDM	0.0100	0.0025	0.0023	0.0026
IPE	0.0112	0.0017	0.0065	0.0018
Rocchio	0.0030	0.0036	0.0034	0.0074
SVM	0.0001	0.0005	0.0001	0.0008

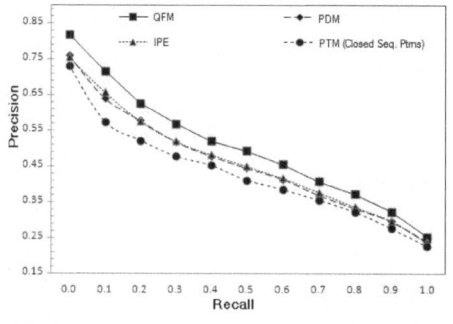
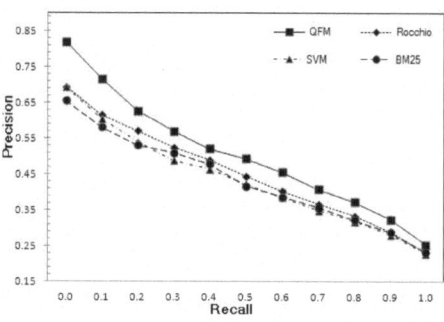

(a) QFM and the pattern-based methods   (b) QFM and the term-based methods

Fig. 3. Comparison Results on precision-Recall Curve

while the average size of query, which is the average number of features (e.g., patterns or terms) used for each assessor topic, $(AVG_{|Q|})$ with the closed patterns is much less than that of non-closed ones. This can be explained that discovering more patterns increases over-fitting and unreliable estimates. Moreover, we found that both PDM and IPE methods are much more effective than all PTM models that use directly patterns across the IR metrics. The results support that both the PDM and IPE methods do perform well the effective use of patterns in text to evaluate term weights.

We also compare QFM with the PDM and IPE methods. As seen in Table 2 and Figure 3(a), the QFM that uses both positive and negative patterns to evaluate the term weights achieved the improvements in comparing with the PDM that uses pure positive patterns. Moreover, As compared to the IPE method, its improvements are consistent and statistically significant (Table 3). Although both the QFM and IPE methods make use of positive and negative patterns, the main reason that the IPE method does not perform well in the assessor topics is that IPE cannot remove noisy terms that causes producing poor queries. Conversely, QFM removes some noisy patterns through analysing a relationship between patterns, where the QFM method uses the smallest number of query terms (102.70 terms) for each topic query (or -83.52% of the average number of document terms) in comparing with the PDM and IPE methods. These results also support that more terms added to a query can be harmful the effectiveness.

5.6   Comparison to Term-Based Methods

In this section, we focus on comparing the performance of QFM to popular state-of-the-art term-based methods. As shown in Table 2, we found that all the term-based methods perform over PTM methods, where the Rocchio method does perform well in comparing with the SVM and the BM25 methods. This illustrates keywords remain the very effective concept for relevance ranking since they have good statistic quality in text. However, the performance results compared between the Rocchio method and the QFM method demonstrate that that the quality of terms extracted from high-level data (sequential patterns) are very effective. since these terms are frequent, and come from specific contexts of document (i.e., sentences and paragraphs) rather than a whole document. Accordingly, the average number of query terms used for each topic query in the QFM method is much less than that of the Rocchio method (i.e., 300 terms) and the other term-based methods.

The QFM method also achieved the excellent performance compared to the SVM and the BM25 methods. The plotting of precision on 11 standard points for the QFM and all the term-based methods on all the assessor topics is illustrated in Fig. 3(b). These results supports the superiority of the QFM method and highlights the effective use of patterns for selecting and weighting relevant terms.

6   Conclusion

In the present paper, we have dealt with the quality issue of features extracted from relevance feedback for query formulation. We proposed a novel relevance feedback method for extracting high-quality features using data mining. We also make use of mining negative relevance feedback to deal with noisy features that cause producing poor queries. Finally, a new term-weighting method for summarizing patterns discovered in text documents was introduced. We promote the proposed method by developing a novel information filtering (IF) model. The experimental results conducted on Reuters Corpus Volume 1 data collection and TREC topics support that the IF model can significantly improve filtering performance in comparing with state-of-the-art IF models. An interesting area for further work focuses on the area of query refinement that finds new query terms for describing the intent of query.

References

1. Bendersky, M., Metzler, D., Croft, W.: Learning concept importance using a weighted dependence model. In: 3rd ACM WSDM International Conf. on Web Search and Data Mining, pp. 31–40 (2010)
2. Buckley, C., Salton, G., Allan, J.: The effect of adding relevance information in a relevance feedback environment. In: ACM SIGIR 17th International Conf., pp. 292–300 (1994)
3. Buckley, C., Voorhees, E.: Evaluating evaluation measure stability. In: 23th ACM SIGIR International Conf. on Research and Development in Information Retrieval, pp. 33–40 (2000)

4. Cao, G., Nie, J., Gao, J., Robertson, S.: Selecting good expansion terms for pseudo-relevance feedback. In: The 31st Annual International ACM SIGIR Conf. on Research and Development in Information Retrieval, pp. 243–250. ACM (2008)
5. Caropreso, M., Matwin, S., Sebastiani, F.: Statistical phrases in automated text categorization. Centre National de la Recherche Scientifique, Paris, France (2000)
6. Carvalho, A., Moura, E., Calado, P.: Using statistical features to find phrasal terms in text collections. Journal of Information and Data Management 1(3), 583 (2010)
7. Croft, W.B., Cronen-Townsend, S., Larvrenko, V.: Relevance feedback and personalization: A language modeling perspective. In: DELOS Workshop: Personalisation and Recommender Systems in Digital Libraries (2001)
8. Dumais, S.: Latent semantic indexing (lsi): Trec-3 report, pp. 219–219. Nist Special Publication SP (1995)
9. Jaillet, S., Laurent, A., Teisseire, M.: Sequential patterns for text categorization. Intelligent Data Analysis 10(3), 199–214 (2006)
10. Joachims, T.: A probabilistic analysis of the rocchio algorithm with tfidf for text categorization. In: 14th ICML International Conf. on Machine Learning, pp. 143–151 (1997)
11. Lam-Adesina, A., Jones, G.: Applying summarization techniques for term selection in relevance feedback. In: The 24th Annual International ACM SIGIR Conf. on Research and Development in Information Retrieval, pp. 1–9. ACM (2001)
12. Lee, C.J., Lin, Y.C., Chen, R.C., Cheng, P.J.: Selecting effective terms for query formulation. Information Retrieval Technology, 168–180 (2009)
13. Lee, K., Croft, W., Allan, J.: A cluster-based resampling method for pseudo-relevance feedback. In: The 31st Annual International ACM SIGIR Conf. on Research and Development in Information Retrieval, pp. 235–242. ACM (2008)
14. Li, Y., Tao, X., Wu, S.-T., Algarni, A.: Mining Specific and General Features in Both Positive and Negative Relevance Feedback. In: TREC 2009 Relevance Feedback Track (2009)
15. Liu, B., Dai, Y., Li, X., Lee, W., Yu, P.: Building text classifiers using positive and unlabeled examples. In: Third IEEE International Conference on Data Mining, ICDM 2003, pp. 179–186. IEEE (2003)
16. Lv, Y., Zhai, C.: Positional relevance model for pseudo-relevance feedback. In: Proceeding of the 33rd International ACM SIGIR Conference on Research and Development in Information Retrieval, pp. 579–586. ACM (2010)
17. Metzler, D., Croft, W.: A markov random field model for term dependencies. In: Proceedings of the 28th Annual International ACM SIGIR Conference on Research and Development in Information Retrieval, pp. 472–479. ACM (2005)
18. Nallapati, R.: Discriminative models for information retrieval. In: The 27th Annual International ACM SIGIR Conf. on Research and Development in Information Retrieval, pp. 64–71. ACM (2004)
19. Nanas, N., Vavalis, M.: A "Bag" or a "Window" of Words for Information Filtering? In: Darzentas, J., Vouros, G.A., Vosinakis, S., Arnellos, A. (eds.) SETN 2008. LNCS (LNAI), vol. 5138, pp. 182–193. Springer, Heidelberg (2008)
20. Pérez-Agüera, J., Araujo, L.: Comparing and combining methods for automatic query expansion. Advances in Natural Language Processing and Applications Research in Computing Science 33, 177–188 (2008)
21. Robertson, S.E., Jones, K.S.: Relevance weighting of search terms. Journal of the American Society for Information science 27(3), 129–146 (1976)
22. Robertson, S.E., Walker, S., Beaulieu, M.: Experimentation as a way of life: Okapi at trec. Information Processing & Management 36(1), 95–108 (2000)

23. Rose, T., Stevenson, M., Whitehead, M.: The reuters corpus volume 1-from yesterday's news to tomorrow's language resources. In: 3th International Conf. on Language Resources and Evaluation, pp. 29–31 (2002)
24. Salton, G., Wong, A., Yang, C.S.: A vector space model for automatic indexing. Communications of the ACM 18(11), 613–620 (1975)
25. Schiaffino, S., Amandi, A.: Intelligent user profiling. Artificial Intelligence, 193–216 (2009)
26. Scott, S., Matwin, S.: Feature engineering for text classification. In: The 16th ICML International Conf. on Machine Learning, pp. 379–388 (1999)
27. Sebastiani, F.: Machine learning in automated text categorization. ACM Computing Surveys 34(1), 1–47 (2002)
28. Soboroff, I., Robertson, S.: Building a filtering test collection for trec 2002. In: The 26th Annual International ACM SIGIR Conf. on Research and Development in Informaion Retrieval, p. 250. ACM (2003)
29. Tesar, R., Strnad, V., Jezek, K., Poesio, M.: Extending the single words-based document model: a comparison of bigrams and 2-itemsets. In: Proceedings of the 2006 ACM Symposium on Document Engineering, pp. 138–146. ACM (2006)
30. Van Rijsbergen, C., Harper, D., Porter, M.: The selection of good search terms. Information Processing & Management 17(2), 77–91 (1981)
31. Wu, S., Li, Y., Xu, Y.: Deploying approaches for pattern refinement in text mining. In: 6th IEEE ICDM International Conf. on Data Mining, pp. 1157–1161 (2006)
32. Wu, S., Li, Y., Xu, Y., Pham, B., Chen, P.: Automatic pattern-taxonomy extraction for web mining. In: 3th IEEE/WIC/ACM WI International Conf. on Web Intelligence, pp. 242–248 (2004)
33. Yanagimoto, H., Omatu, S.: Information filtering using a probabilistic model. Artificial Life and Robotics 10(1), 41–44 (2006)
34. Zhong, N., Li, Y., Wu, S.: Effective pattern discovery for text mining. IEEE Transactions on Knowledge and Data Engineering (99), 1–1 (2012)

Author Index